THINGS FOR CHILDREN
TO MAKE AND DO

By the same author

SCIENCE FOR HANDICAPPED CHILDREN

HUMAN HORIZONS SERIES

THINGS FOR CHILDREN TO MAKE AND DO

Science and Technology Activities

Alan V. Jones

A CONDOR BOOK
SOUVENIR PRESS (E & A) LTD

ACKNOWLEDGEMENTS

A large proportion of this book has arisen from a project guided by the Science Education group in the School of Science at Trent Polytechnic. Entitled School Science Research Project (SSRP), it has been funded by major contributions from Manpower Services and Trent Polytechnic Research Fund, together with valuable donations by Boots, BP, British Telecom, and NatWest, Barclays, Lloyds and Midland Banks. Without the help of the above, the project would not have been able to employ the teachers and researchers needed to develop and test in schools this and other material.

The major contributors have been lecturers, teachers, advisers/inspectors, and researchers, and include the following:

    John Carter
    Alan Jones
    Jim McLaren
    Cyril Gilbert
    Charles O'Brian
    Colin Hughes
    David Coleman
    Susan Tomlinson
    Paddy Rayner
    Richard Wheat
    Ron Hinton.

Members of the SSRP, including:

    Richard Middleton
    Linda Evans
    Alan Sanderson
    Chris Richards
    Robert Brown
    Maureen Morby
    Anne Topping.

Thanks are also due to the Nottinghamshire schools and the teachers who tried the experiments and gave valuable evaluation of the work along with comments, analysis and constructive criticisms. Teachers on 1/86 DES courses (Science and Technology for teachers of pupils with special educational needs) also gave valuable advice.

Typing and word processing were done by Maureen Morby.

Diagrams and artwork are by Richard Middleton.

This book is a continuation of the ideas and activities given in an earlier book, *Science for Handicapped Children*, also published by Souvenir Press (1983).

CONTENTS

PREFACE

This collection of ideas, involving science, technology, design, problem solving and construction, links quite naturally with the ideas given in a previous book, *Science for Handicapped Children.*

The activities chosen for study here have focused on the everyday things around us and an investigatory approach has often been used.

The word technology as used in this text means the applying of scientific and commonsense principles to help to do useful work and solve practical problems.

The parts of the book have been chosen to be of as wide a use as possible, that is, to parents, teachers and pupils. The first part, 'Starting Science and Technology' (using experiments and investigations), deals with introductory activities, either for pupils just starting science, or for older pupils whose abilities are more suitably matched to the requirements of simple, teacher directed experiments.

The second part uses a problem solving approach to capture the imagination of pupils (and adults) of all ages, and there are items which positively encourage the pupils' creativity and design skills.

The third part covers those aspects of science and technology which would help the parents and teachers, and includes some case studies and ideas of other practising teachers.

INTRODUCTION

Since the full realisation of the implications of the 1981 Education Act in Britain, which allowed for integration of children with special educational needs into mainstream classes, the interest in topics of science and technology for all children has increased. More children have had access to science activities in mainstream schools and even those children remaining in special schools and hospitals have had their school curriculum more scientifically enriched. The development of science based toys, constructional and technological apparatus, the use of the computer—all these have influenced everyone's lives. A report of the many visits made by Her Majesty's Inspectors of Schools to a variety of schools found that there was a need for more science and technology to be taught within the curriculum of all pupils with special educational needs. The report went on to say that there was also a need for more in-service training of teachers to encourage them to make use of the motivating ideas included in scientific and technological projects.

This book cannot fully do justice to *all* the aspects of science and technology, but it is hoped that it will open up *some* areas which the children will not have used in the past.

The book is primarily for parents and teachers, or for pupils who have the reading ability to follow experimental details. Because every child is an individual, it is realised that it is impossible to write a completely satisfactory set of work instructions for each pupil. It has thus been decided to write the book in simple English and with the minimum of jargon, for teachers and parents to read to the pupils and explain the procedure as required. Many will use the book as a source of ideas and many will find the work-sheets suitable for use with the pupils in their class; for other pupils the details might need modifying slightly.

The approach adopted for the activities
Wherever possible the approach has been a practical and investigatory one, emphasising the following items:
1   To encourage the pupils to do the experiments themselves.
2   To encourage a sense of achievement and feeling of independence by using the experimental approach.
3   To help the pupils to ask questions, think and communicate either verbally, in writing or by drawings.
4   To provide safe, interesting and relevant activities, usually on a particular theme.
5   Everyday material is used and laboratories are not needed, but a sense of doing real experiments must be maintained and encouraged.
6   The experiments can be done in schools, at home or in hospitals.

7   The experiments do not usually require the pupils to have much previous knowledge of science and each activity is designed to make them think about the present investigation, and to solve relevant, meaningful and interesting problems.

Who can do the activities?

Usually, if the children have the physical ability to feed themselves, or the mental ability to understand simple instructions, then they can cope with the individual activities outlined here. If this is not the case then group activities might be a suitable answer, or the help of a parent, teacher or voluntary aid. It is essential that even though some help is given, opportunity must be given for the pupils to think, observe and try out the ideas for *themselves*.

The apparatus needed can usually be found around the home and care has been taken not to include obviously unsafe materials. It is, therefore, IMPORTANT to:

1   follow instructions;
2   never use mains electricity for the experiments;
3   be careful when handling hot water or lighted candles, sharp instruments, scissors and knives.

It has been generally found that physically handicapped people are as safe workers as anyone else, probably because they know their limitations more clearly. For all special needs pupils the practical approach to experimentation has been shown by many to be a means of increasing the children's confidence and ability to prepare themselves for independent living and practise real life skills.

The activities are designed for children who are beginning to organise objects and apparatus constructively to achieve their goals, and who are beginning to see the connection between cause and effect (i.e. they can see the connection between flipping a switch and a light going on). This stage is characterised by asking suitably guided questions and posing simple, single-stage problems. As children develop they will begin to want to investigate their surroundings in a constructive and controlled way. They will want to know the answers to experiments and pose further ones, and will eventually be able to construct a logical method of solving problems. It is probable that not all pupils will reach the higher stages of these activities but all should be able to investigate some of the problems posed.

Some pupils are good thinkers but slow manipulators or have dexterity problems. In such circumstances group work or helpers may be needed.

Pupils who will benefit from the various activities in this book may suffer from any of the following disabilities (or others):

Very slow learners	Cerebal Palsy (Spastics)
Hearing difficulties	Spina Bifida
Visually impaired	Muscular dystrophy
Learning difficulties	Orthopaedic handicaps

12

The book will also be useful to pupils who are patients in hospitals, and it has been found that the activities have given a motivating challenge to adults who have suffered from paralysis or partial strokes and who are house-bound or attend day centres.

For parents and teachers

This book supplies some ideas and tested experiments, but on their own these cannot provide the vital ingredients for success in scientific investigations, namely motivation and enthusiasm. Even though I am enthusiastic as I write, I am not necessarily conveying this to the child. This is where your expertise is needed.

If you have used any of the experiments in *Science for Handicapped Children*, you will already be convinced of the joy and motivation the pupils get out of doing experiments themselves.

The intentions of this book

It is *not* intended to be a complete encyclopaedia of every activity in science or technology that can be taught, neither is it a complete 'syllabus' in itself. It does, however, set out to show how, with simple apparatus, an investigatory approach to science and technology can be used with children of any age and ability. The application of the scientific principles can be of use not only to them but to the world at large.

The initial investigations can be carried out with children at any age, but the answers to the questions posed and problems set will be dependent on their abilities and experiences.

It has been found that many pupils have not been allowed to play with junk or mess around with wheels, cogs or chains, and so have often been deprived of the opportunity to find out about various mechanisms, thus accumulating worthwhile experiences. Investigations are, therefore, a new dimension to some, particularly those who have been over-protected and have not needed to think for themselves and do things themselves. Often a little bit of encouragement to the pupils to 'get stuck into the task' can help them produce worthwhile solutions to problems.

Apparatus

One of the main aspects of the investigatory approach at home, or with limited apparatus or laboratories, is to collect any item which looks useful, e.g. plastic containers, tops of bottles, coffee jars, screw-top jars, tins, cardboard of various thicknesses, etc.

This simple apparatus does not in any way detract from the quality of science and technology that can be undertaken.

Language and life skills

It has been found that, as a result of doing active and interesting experiments, the children want to tell someone about their findings; this can be done in as many ways as possible, from posters, scrap-books, sound or video tapes, to talking and showing others their activities.

Science and technology experiments can be an ally to teachers and parents trying to encourage pupils to improve their communication skills.

Because many special needs pupils lead a sheltered life, protected from life's tasks (cooking, mending things, etc.) by adults and helpers, they may lose the feeling of wanting to solve any problems. These children have often become much more passive than need be. When they leave school or college they will be faced with a multitude of technology surrounding them. They must therefore be encouraged to participate while they are young, to get involved in doing active science and technology.

Many children and adults have a fear of mathematical calculations and because of this the activities have been chosen with little or no mathematical bias.

Things to remember when doing the experiments
It is necessary, when doing any of the investigations, to encourage the pupils to talk about their ideas and to express them in words, diagrams or writing. Talking and explaining can help the pupils to understand the related concepts.

Sometimes experiments do not work the first time, or they need modifying slightly to get them to work. This is what science, technology and problem solving is all about. Sometimes you can think of a better way of solving the problem than the one given in the book . . . great . . . go on and try and see if it works. That's how discoveries have been made.

Skills and processes
The whole collection of activities is designed to encourage the development of a series of practical, manipulative, thinking and reasoning skills. These include the encouraging of:
> observations,
> following instructions accurately,
> making inferences and deriving principles and conclusions,
> measuring,
> making predictions and testing them,
> constructive and manipulative skills,
> communication—verbal, written, pictorial,
> stating ideas and theories and testing them,
> using knowledge and practical skills to solve problems.

Some notes before you begin
This collection of activities is designed for people who would like to *do* some investigations in the home or school.
1   They are designed for any person who has the ability and mobility to hand-feed themselves. We have kept in mind an orthopaedically handicapped person and also a blind person who will be able to do most of the experiments, depending upon the degree of vision.

14

2   The experiments can be done by anyone (of any age), but some basic level of reading and ability to carry out instructions is assumed. It would be possible for a second person to read out the instructions and interpret them to the experimenter and still get a lot of scientific enjoyment out of the experiments.

3   The laboratory is right beside you—your classroom, the living room and kitchen. All are ideal for scientific investigations, and the 'apparatus' we are going to use can easily be found around the home or purchased in a local shop.

   At the start of each experiment there is a list of the things you need to carry it out.

4   All the experiments are as safe as we can make them and there is no chance of you being blown up doing the chemistry experiments or electrocuted using the 4·5 volt batteries. Any safety precautions needed are given in the details of the individual experiments.

5   Often, the first time you do an experiment you are just getting the feel of the apparatus, and sometimes you might have to repeat it a second time to get satisfactory results. All scientists and engineers have to do this.

6   Many scientists and engineers have started their work by doing home experiments.

7   We have written this book so that you can take an experiment from any part of the book, but it is best to work through one section at a time.

8   Often one experiment leads to another, and *if you do your own experiments*, not in this book or other books, please make sure they have been checked for safety.

IMPORTANT SAFETY HINTS

Never use mains electricity for experiments.
Keep water and wet hands away from mains electrical plugs and lights.
Be careful when handling hot or boiling water or lighted candles, etc.
Take care using sharp instruments, knives, scissors, pins and needles.
Enjoy your experimenting.

How to use this book of activities
It is probable that, because you have picked up this book, practical activities interest you. The activities can be used in any of the following ways.

   1   As a collection of activities to be used sequentially and periodically (e.g. two hours per week).

   2   To supplement or enrich any relevant topics, e.g. transport or flight, etc.

   3   As a problem solving session.

   4   As the basis for a constructional project.

   5   In special schools or with pupils in special units or pupils in hospital schools.

   6   To encourage a feeling of importance and independence.

   7   To encourage group activities and group co-operation.

The need for flexibility
The educational requirements of children with special educational needs are much more individual than those of children in the mainstream of schools.

The ideas included in this book have often been omitted from the curriculum of pupils with special needs, but for anyone using the activities they have been seen to be a great motivational force and a further way to encourage meaningful communication. They also give the pupils an opportunity to be successful and creative: this is important as many pupils have become uncreative and unsuccessful in other curriculum areas and may sometimes become passive observers instead of active participators.

The instructions can be read to the child or put on sound tape for a group of pupils. The findings of an experiment or activity need to be recorded for future reference and these can be written or put on a sound tape, etc.

All the activities included in this text have been developed and tested in schools with a wide range of pupils.

PART 1

STARTING SCIENCE AND TECHNOLOGY

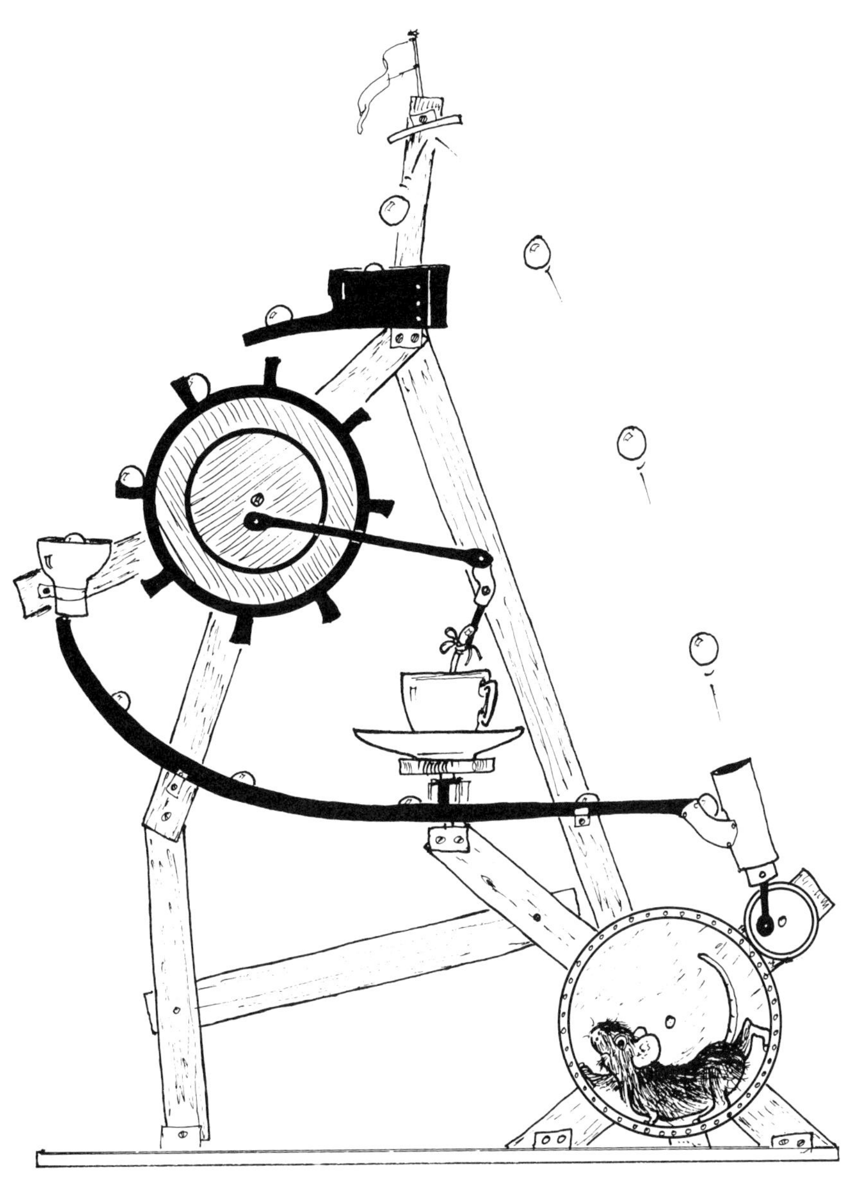

CONTENTS, PART 1

About Part 1

The activities in this part are arranged to motivate pupils and guide them carefully through a number of useful and interesting ideas. The experiments often give basic principles which will be extended in Part 2, which tends to offer more open ended activities. These allow the pupils to apply the principles of Part 1 and seek to allow the pupils to think in a creative manner and thus help to solve the problems set.

EXPERIMENTS WITH MIRRORS

Information for Teachers and Parents

Skills and Processes Included in this Section
It is hoped that by doing the experiments in this section the following skills will
be practised and so improved :
    Drawing
    Measuring lengths and measuring angles
    Observing, counting and investigating
    Predicting and thinking
    Constructing and manipulating apparatus
    See the applications to everyday life, use of reflections and mirror reversal

Words and Vocabulary Needing Explanation
Reflection, opposites
Mirror images
Symmetrical
Angles

Materials Needed
For the section on mirrors you will need in total :
    Two (or three) small mirrors (4 cm × 6 cm approx.)
    Few sheets of A4 white paper
    Piece of Blu-tack
    Pen or pencil
    Some stiff card (for making a periscope) e.g. cornflake package

Experiments with Mirrors
This collection of experiments is essentially trying out ideas, observing and inter-preting results. No explanation of ray diagrams of light paths is attempted as it is thought to be inappropriate and abstract, but a few guided questions will give ideas to the pupils to help them understand what is happening in the light, reflections, etc.

Questions to Think About at the Start of this Topic
Stand or sit in front of a mirror and pick up a comb with your right hand. Is the reflection in the mirror showing the comb on the left or right hand side?

What causes the reflections? Why can I see myself in a mirror but not on the surface of the wall? What other things around the home or classroom show me my reflections? Are all these the same colour, e.g. silver?

Mirrors and Reflections

You will need: A small mirror
                 Writing paper
                 Pen or pencil

What to do:

Look at the drawing of half a house

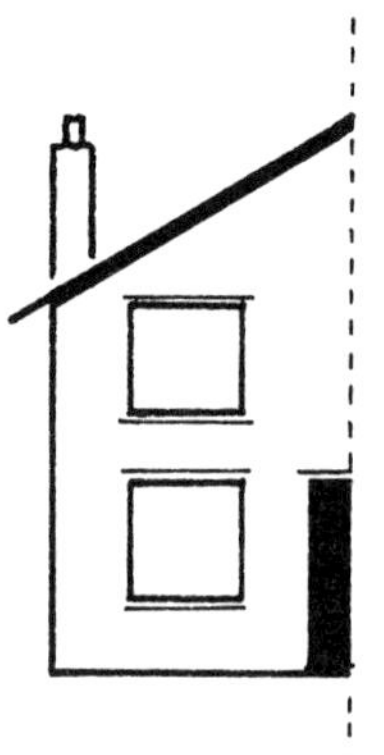

Put your mirror along the dotted line, shiny side facing the house, and look into the mirror.
Have you now got a picture of a complete house?

How many windows does your complete house have?
How many doors?
How many chimneys?
We say when both halves of a picture are identical it is 'symmetrical'.
Is your house symmetrical?

Try these by putting your mirror along the dotted line.

How many letters of the alphabet are symmetrical? (Try them with your mirror.)

Look at a photograph in the newspaper or look at your own face. Is your, or anyone else's face perfectly symmetrical?

Mirrors and Angles 1

EXPERIMENT 1

You will need: Two small mirrors
Sellotape
Blu-tack

What to do:

Join two mirrors, shiny sides facing each other, at one edge with sellotape.

Stand the mirrors on the lines below and put a small pencil on a piece of Blu-tack on the circular mark. Look into the mirrors and count how many pencils you can see.

Number of pencils seen   Number of pencils seen

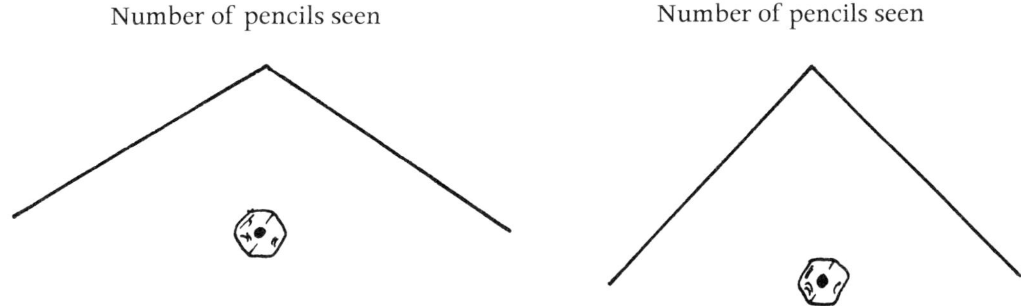

Number of pencils seen

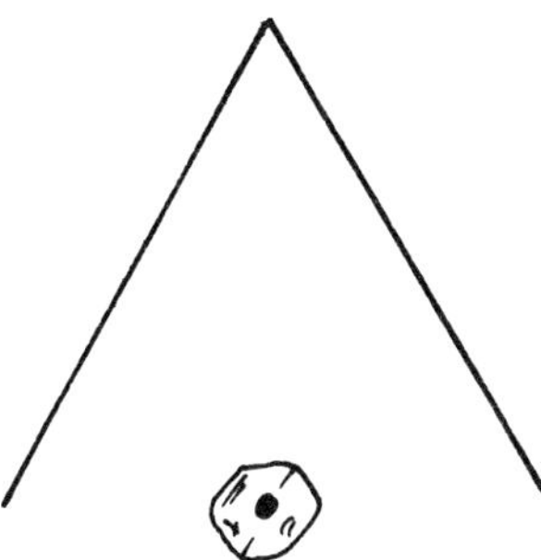

Does it make any difference where you put the pencils to the number of pencils you can see?

What would happen if you joined three mirrors together and made shapes like
the ones shown below? How many pencils would you see then?

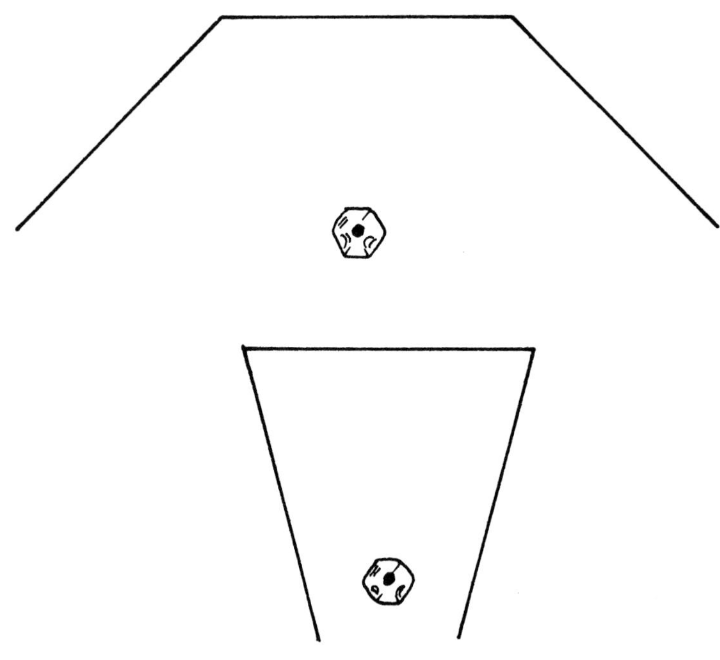

Suppose you used
two different-coloured
pencils at different positions.
What would you see?
Like to try it?

Could you arrange your mirrors to help someone see the back of their head?
(This could help them combing their hair!)

Mirrors and Angles 2

You will need :   2 mirrors
Blu-tack
Pencil
Few sheets of stiff card (e.g. cornflakes packet)
Few sheets of paper
Sellotape, scissors, matchstick (dead)
45° set square or protractor

The following experiments are going to show how useful mirrors can be to people.

List some of the uses of mirrors to people, have a look around the home, classroom and outside. Note how often we use reflections or mirrors to our advantage.

Try these experiments

EXPERIMENT 2

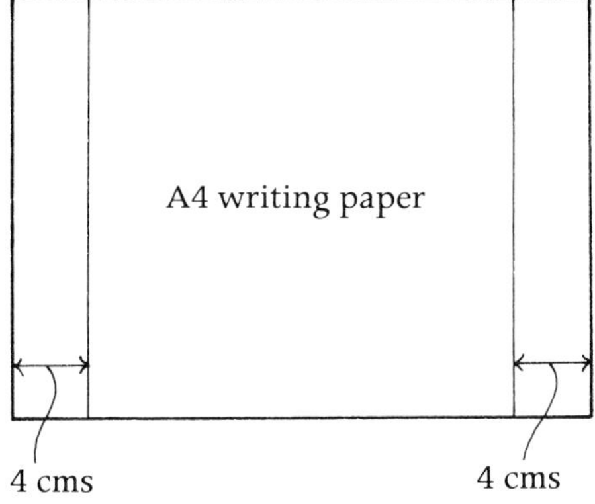

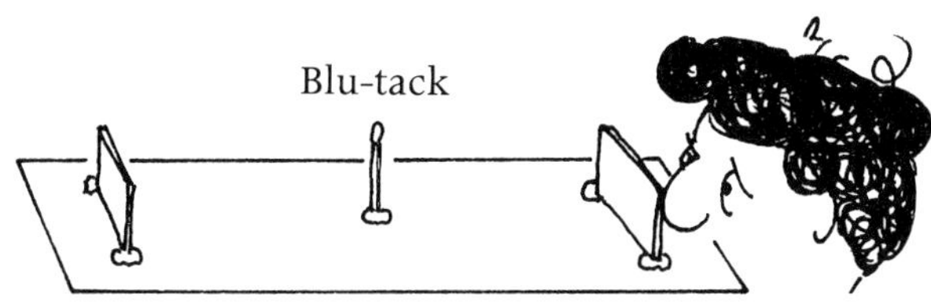

1   Draw two lines 4 cms from each end of the A4 paper.

2   Put the paper flat on the table and Blu-tack the two mirrors standing up facing each other on the two lines.

3   Put a matchstick on Blu-tack between the two mirrors.

4   Look over the top of one of the mirrors and look at the other mirror.

How many matchstick reflections can you see?

You might need to move the mirrors slightly to see what difference that makes.

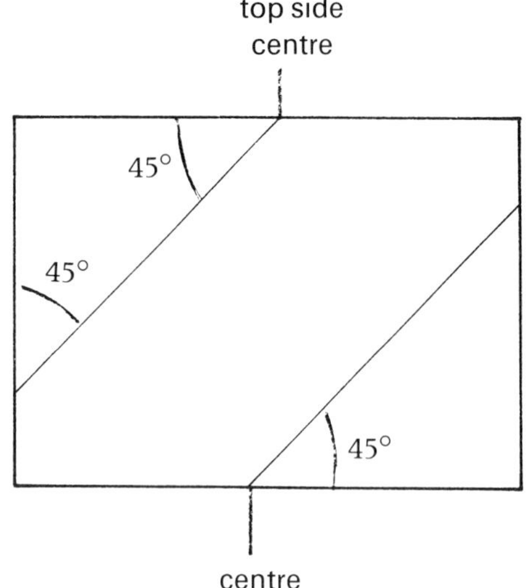

1 With the same sheet of paper mark the centre of the lines and measure 45°; draw a line across paper as shown.

2 Now put the mirrors facing each other with their centres at the centre of the lines.

3 Looking only in the direction shown, where can you put the matchstick so that it can be seen in mirror A? Try various positions. How can you see it in A? Has the angle of mirror B anything to do with it? Move the angle of mirror B and experiment.

Extension of Experiment 3
Do the angles always have to be at 45°? Do they always have to be parallel?
If you set mirror A at 30°, what would you have to do to mirror B to see the matchstick along the top side by only looking into mirror A as before?

EXPERIMENT 4

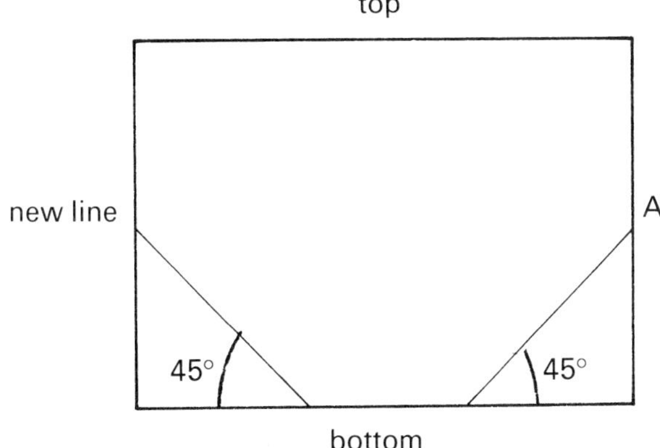

1   Leave mirror A along the 45° line and draw a 45° line in the direction shown.

2   Put the two mirrors facing each other now.

3   By looking in mirror A can I ever see the matchstick placed anywhere along the top line? Predict and then try.

4   Think first and predict if I could see the matchstick in mirror A along any position along the bottom line. You might have to move the mirror slightly to get them lined up.

5   Has the angle of mirror B anything to do with your ability to see the matchstick on the bottom line? Move B slightly and see.

Can you see if any of the mirror positions are used in everyday life?

If you wanted to see the back of your head using two mirrors, what arrangement would you use?

If you wanted to see around a corner, what arrangement would you use? Who might want to see if any danger was awaiting him around the corner?

If you wanted to see under a car to see if there was a bomb attached to the underneath of the car, how could you use a mirror to do this?

How does a submarine under water use mirrors to see ships above the water level?

How does a dentist use a mirror to see the back of your teeth?

In a later section of the book some problems will be set using some of the ideas you have tried here.

You can either try these problems now or return to them later.

WHEELS AND THINGS THAT TURN

Information for Teachers and Parents
The extensive use of wheels for movement and turning mechanism often means we take them for granted and miss some of the usefulness of turning wheels.

Skills and Processes Included in this Section
Observing and investigating
Making and manipulating
Thinking and asking questions
Looking for applications

Words and Vocabulary Needing Explanation
Rotation
Opposite movement
Axle, revolutions, revolving

Materials Needed
Piece of thick peg board approximately 30 cm × 30 cm (see below)
Empty cotton reels
Tops from coffee jars, jam jars, pop bottles, etc.
Dowel for peg board
Elastic bands of various sizes
Hammer to knock the nail in Base Board

Base board: 30 cm × 30 cm approx. This can be made of a piece of peg board and dowel pegs can be placed into position as required.

Tips
If, when turning the cotton reels with elastic bands on them, the elastic band slips, then roughen up the reels with a bit of sandpaper and use wide elastic bands.

Wheels and Things That Turn

You will need: Base board (peg board)
Dowel pegs
Cotton reels
Coffee jar lids
Elastic bands of varying sizes (they can be tied together to make long ones if required).

What to do:

Set up the pegs in the peg board base so that the two cotton reels, when put on the pegs, are about four centimetres apart.

Connect the two reels with a suitable elastic band so that when one cotton reel turns the other will also. Put two arrows on the cotton reels either by sticking them on or by drawing them on with felt tip pen.

Turn cotton reel A round once.
How many times does B go round?
Does B move in the same direction as A?

Now cross over the elastic band and turn A once. Does B now move in the same direction as A?

Now try this arrangement.

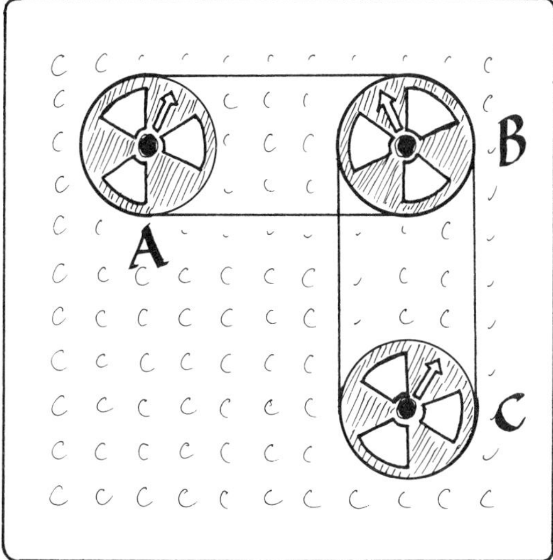

Will C move round in same direction as A?

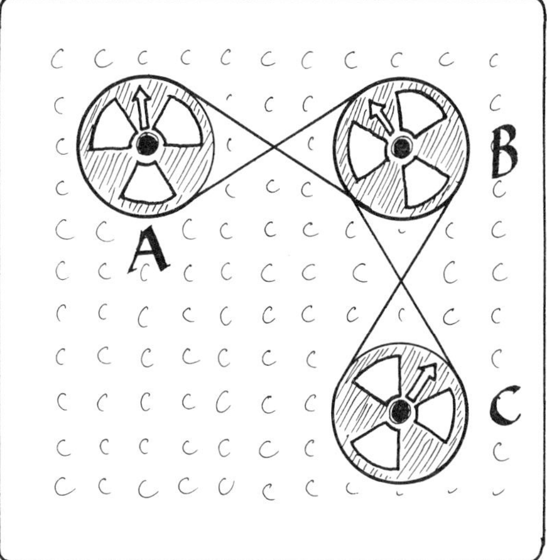

What will happen to C now?

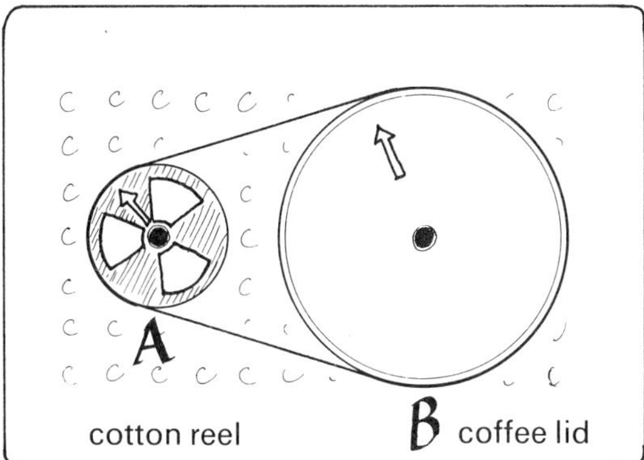

Will B move round more
or less for one complete
turn of A?

Touching Wheels

You will need: Base board
                       Pegs
                       Cotton reels
                       Coffee jar lids or other containers

What do do:

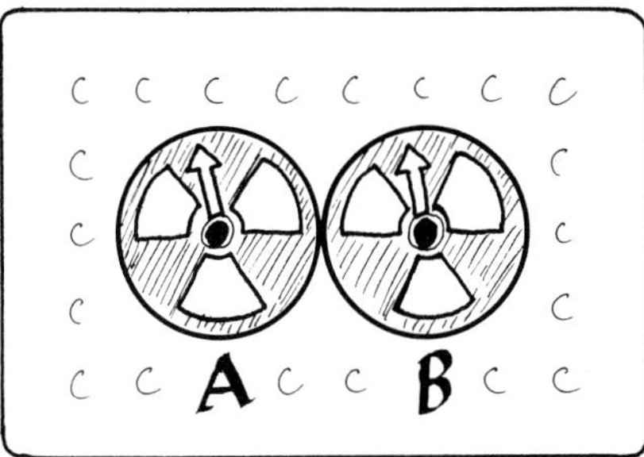

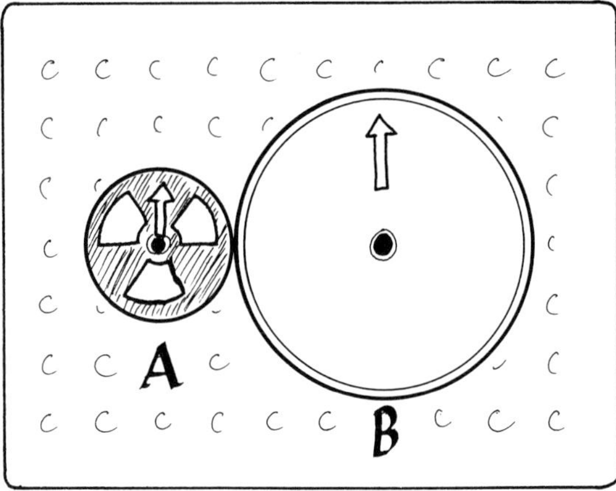

Place a wheel B (cotton reel or coffee jar top) just touching cotton reel A. What happens to B if A is moved round once in a clockwise direction? (*No elastic bands* between them.) What will happen to *a larger* wheel if it is made to just touch wheel A?

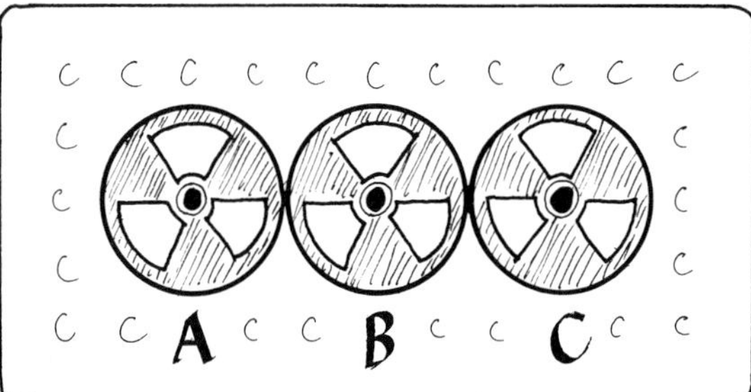

What will happen to a third wheel if placed in contact with B and then A turned?

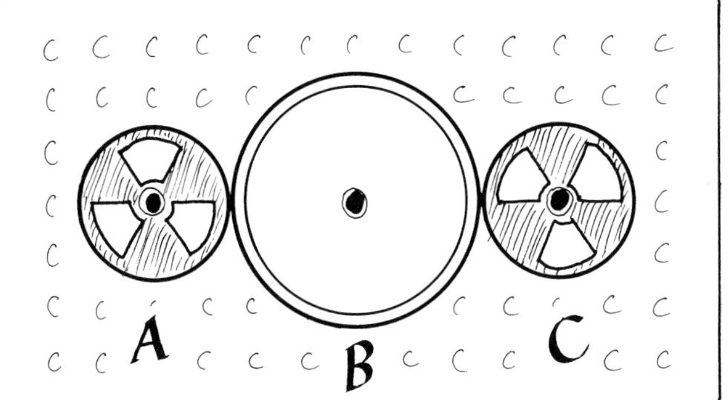

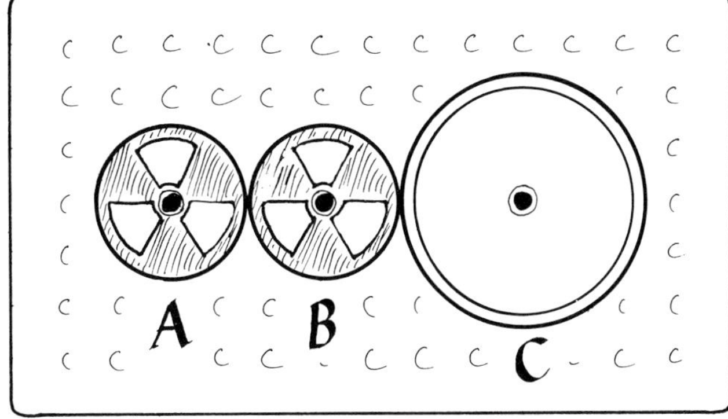

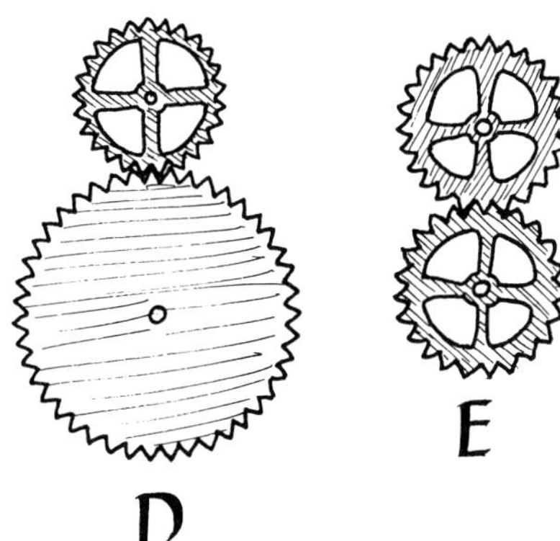

Can you see the usefulness of gear systems for changing the direction of things or changing the speed of rotation of things? A clock uses a mechanism like this to alter the speed of turning of the hour, minute and second hands. What will happen to the other wheel when D or E is turned round once?

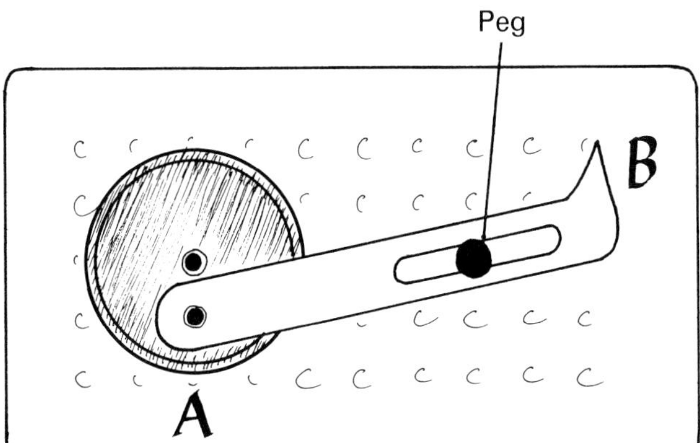

Cut a piece of stiff paper or thin card to form this shape:

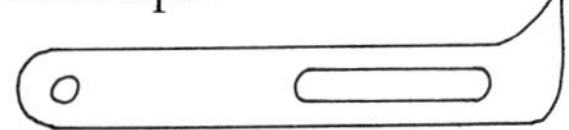

What happens to the pointer on B when A is turned round a few times? What happens to wheel A *if* B is moved backwards and forwards a few times?

Wobbly Wheels

You will need: Two wheels, say from coffee jar lids, one of which has an axle
hole *NOT* in the centre
Elastic band

Try this:

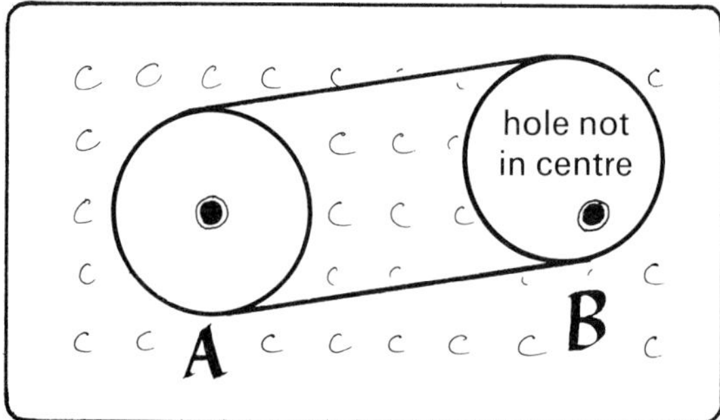

When wheel A is turned round once,
what happens to wheel B?

What would happen to the movement of the car if it had wheels like B?

Have you every used a Spirograph for drawing patterns? Ask if there is one
available in school and try it. It uses the principle of off-centre axle for drawing
shapes.

Can you see a use for the machines you have made?

Take a look at the bicycle, for example.

Predict what happens to the bigger wheel of the bike if you turn the pedals once. (Hint: look at the size of the cog wheel attached to the back wheel)

A problem to think about (and a model to make)
A large windmill is used to turn a wheel (A) inside the mill. Wheel A is connected to wheel B with a drive belt. What happens if the windmill moves slowly round once? Will wheel B move faster or slower than A?

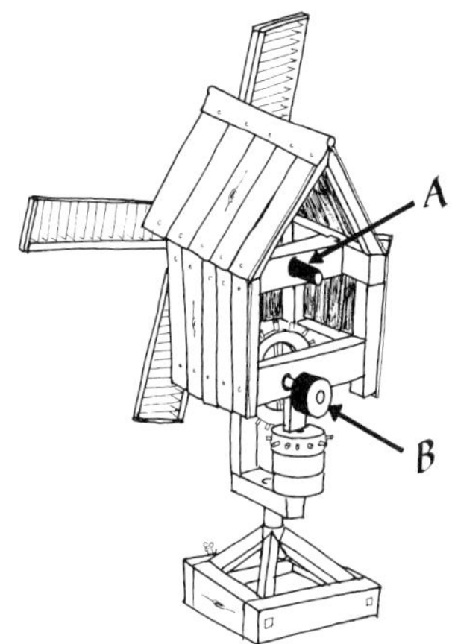

To think about
Do both spools of a cassette tape recorder (or video recorder) move in the same direction?

Some interesting models and problems will be given to you in the next section of the book which will use some of these important principles. You can try these problems now or later.

ELASTIC POWER

The power that is stored up by twisting or stretching an elastic band can be turned to some use if controlled. The following experiments use some of the power stored in elastic bands.

Information for Teachers and Parents

Skills and Processes Included in this Section
Constructing and manipulating
Thinking and predicting
Measuring

Words and Vocabulary Needing Explanation
Expanding and stretching
Twisting and turning
Power from releasing elastic bands

Materials Needed
Peg board (offcuts) about 10 cm × 6 cm
Clothes peg
Elastic band
Dowel
Cotton reel
Candle
Pencil
Empy washing up liquid bottle
Wire clothes hanger

Table Football or Cricket Players

Materials: Peg board approximately 10 cm × 6 cm
Clothes peg
Dowel to fit the peg board
Elastic band
Marble

What to do:

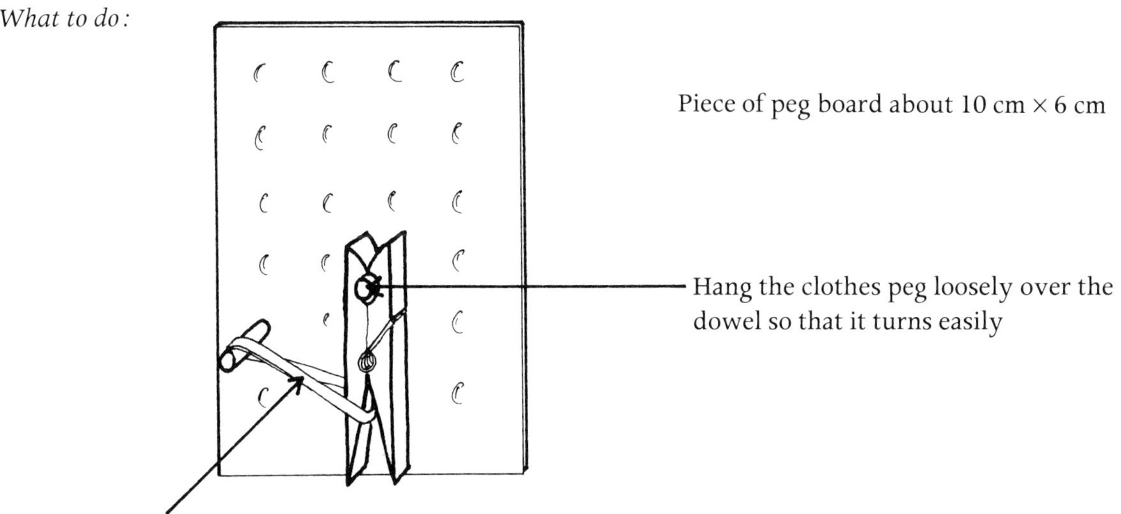

Thread elastic band over the end of
the clothes peg and around the dowel

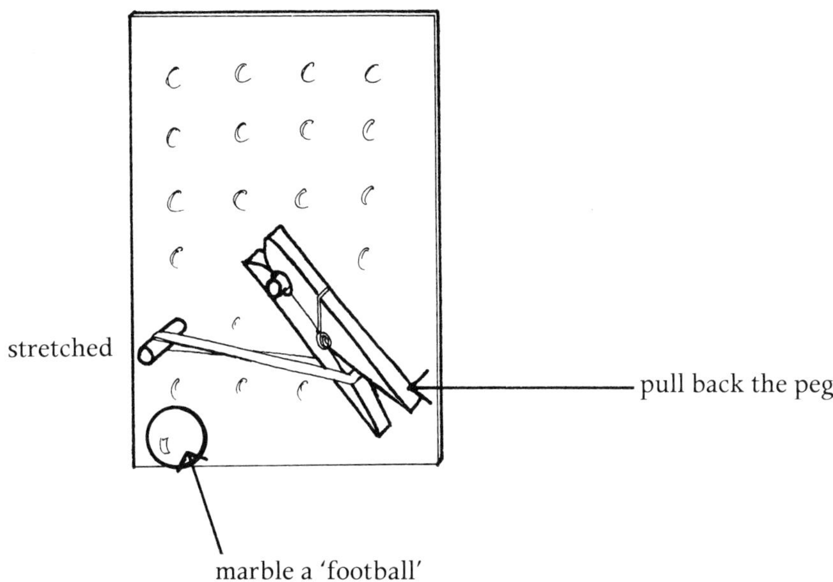

On releasing the held back peg the 'ball' can be kicked.

38

Extension
An artistic person can see how to make this machine look like a footballer.

This leg can be coloured or painted

Can you make a cricket player with a good throwing arm? Or a hockey player with a moving hockey stick?

You might like to make some goals to shoot the ball at, or perhaps a set of wickets and a player with a moving cricket bat also.

How is power in the elastic band similar to the power of a person's muscles?

Put your hand on one of your muscles and see what happens to it when you use your arm or leg.

Shooting Machine

Materials: Spring clothes peg
2p or 10p coins
Elastic band

What to do:

Lift the spring and thread elastic band under the wires.

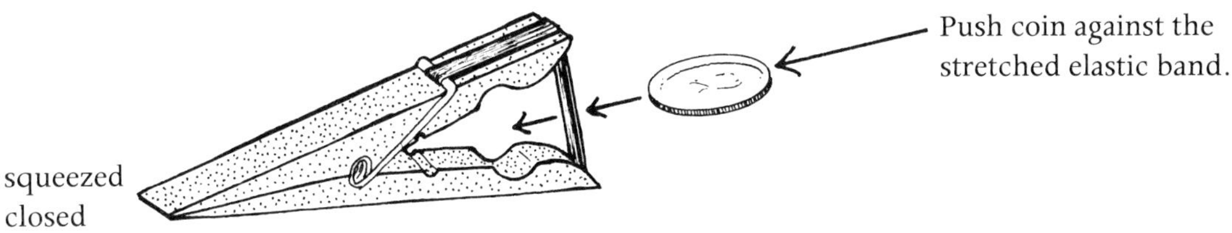

Push coin against the stretched elastic band.

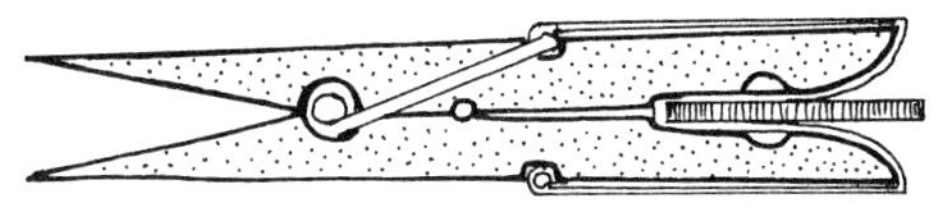

Squeezed closed with stretched elastic band under the coin, now held by the jaws of the peg.
By squeezing the other end of the peg you will release the coin.

You will have to experiment to discover the best elastic band to use.

Be careful where you aim this shooting machine.

Devise a game using your peg gun to shoot coins at containers or yoghurt cups and margarine containers.

Wooden or plastic pegs work equally well.

Crawlers and Tanks

Materials: Elastic band
Match stick (dead)
Pencil
Cotton reel
Candle
Knife
Drawing pin

What to do:

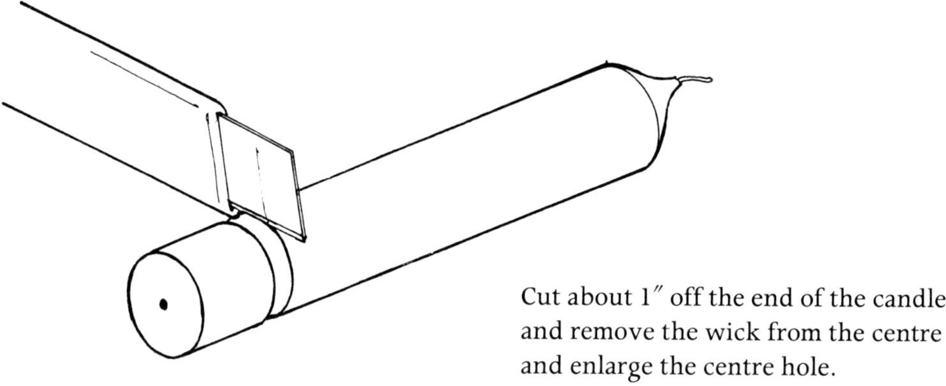

Cut about 1″ off the end of the candle and remove the wick from the centre and enlarge the centre hole.

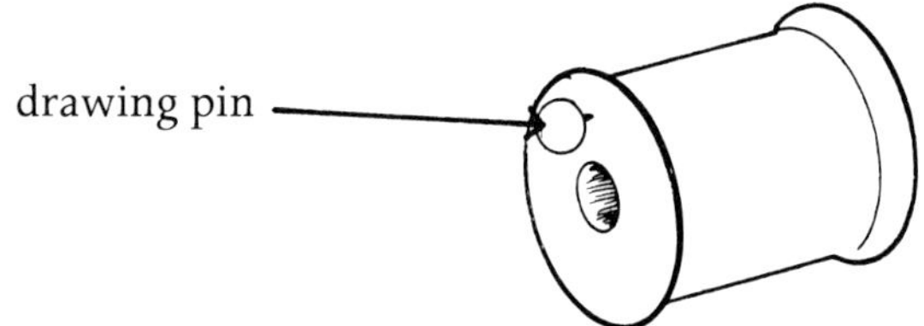

The drawing pin is used to keep the match stick from moving, or stick it down with Sellotape (see next diagram).

Place the end of the candle against the cotton reel.

Thread the elastic band through the holes in the cotton reel and candle.

Thread over the matchstick on the cotton reel end and over a pencil at the candle end.

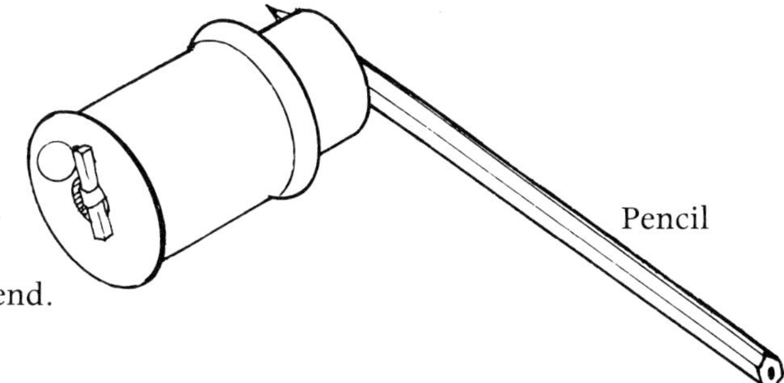

Wind the pencil around as many times as will just allow the pencil and candle to turn slowly. Place the 'tank' on a smooth floor.

Questions
How many twists allow the tank to move?
Do more twists make it go further?
Why is the piece of candle used?
Will it move without the candle?
Will your tank climb up slopes?
What is the steepest slope it will climb?
Does your tank work best on smooth floors or carpet?
If you wanted your tank to climb up a steep, smooth slope what would you have to do with the surface of the tank?

Extension Experiment: Jumbo Size Tank

Materials: Washing up liquid bottle
Elastic band
Pencil
Match stick

Construct in a similar way to the previous experiment.

You might need a piece of wire to help thread the elastic band through the centre of the bottle.
Loosen the top of the bottle.
Suppose you put pencils as 'pushers' at both ends.

Try other containers as possible tanks. Notice that the power of the twisted elastic band is released slowly using friction from either a candle or the cotton reel, etc.

How can you increase the friction of the rollers to allow it to climb up smooth slopes?

Roundabouts
If, instead of moving along the floor, the machines were stood on end, could the power of the elastic band be used for turning things like roundabouts? Design and make one.

Some later problems will use this principle.

Information for Teachers and Parents

Skills and Processes Included in this Section
Drawing
Thinking and predicting
Reading and asking questions

Words and Vocabulary Needing Explanation
Gravity
Force
Mass and weight
Weightlessness
Space

Materials Needed
A pencil (or anything to write with)
An apple (if possible)
A model car plus two pieces of wood to make slopes. (Use wood twice the width of
    the car and about a metre long)
Two plastic cups and some water
A plasticine model of the earth with model 'people' to be stuck on it
A small ball (the apple will do)
A model of a Saturn V rocket—as made by 'Airfix' (but not essential)
A ball-bearing and a ball of plasticine the same size

What is Gravity?

There is a story that about 300 years ago a man called Isaac Newton was sitting under an apple tree. Suddenly one of the apples fell to the ground. As he watched the apple he asked himself: why does the apple fall down and not up?

From the earliest times objects have fallen towards the ground and not towards the sky. We take it for granted that things will fall down and not up, but have you ever wondered why this is so?

Ask yourself: why does the apple fall down and not up?

Take an apple and hold it in your hand. (The nearest object will do if you do not have an apple.)

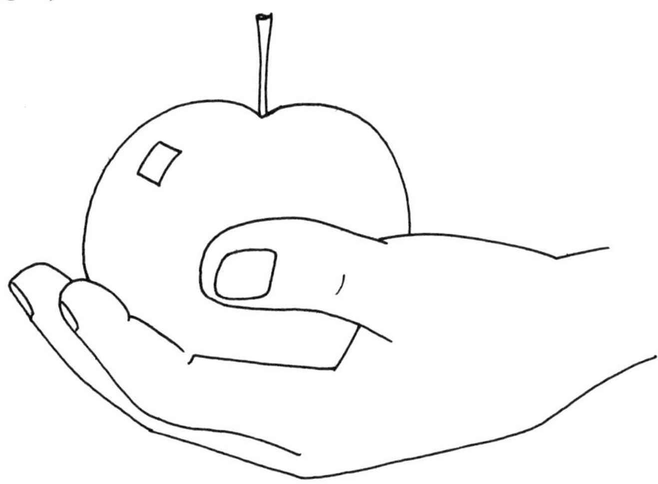

Can you feel the weight of the apple on your hand?

Draw an arrow on the last diagram to show which way the apple would go if you took your hand away.

You have now drawn a line showing the direction in which the earth pulls the apple towards it.

The reason why things fall down to earth and do not go up towards the sky is due to the force of *gravity*.

Gravity is the force which pulls things towards the earth.

We can see the signs of gravity everywhere.

It is easier to bike downhill than uphill.

Water goes down the waterfall and not up it.

Draw arrows on the diagrams to show the direction of gravity.

Make a couple of slopes with some pieces of wood.
Push a trolley, or toy car, up and down the slopes.

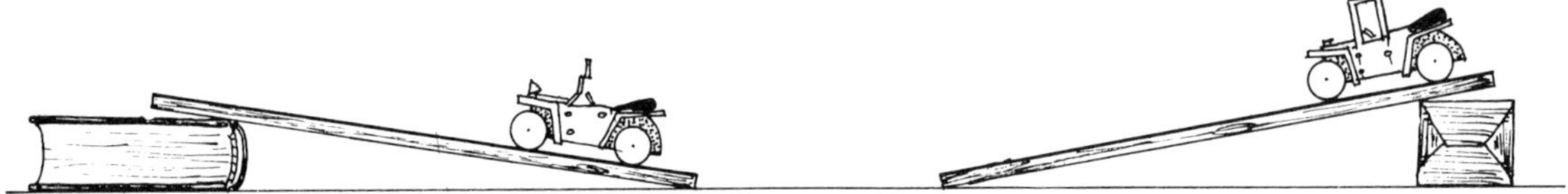

In which direction is it easiest to move the car?

When the car moves down the slope gravity helps to pull it down. But when the car moves up the slope then gravity pulls it back and slows the car down.

Gravity and Our Earth

Many years ago men thought that the earth was flat. This is not true, the earth is a sphere moving through space. So why do Australians below us not fall off?

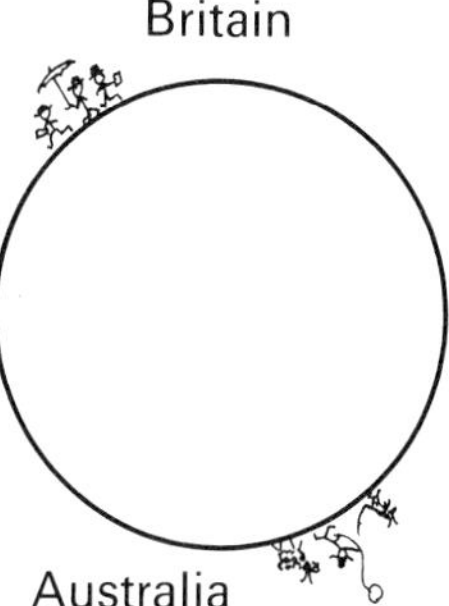

Make a small plasticine model of the earth. People live all over the planet but they do not fall off.

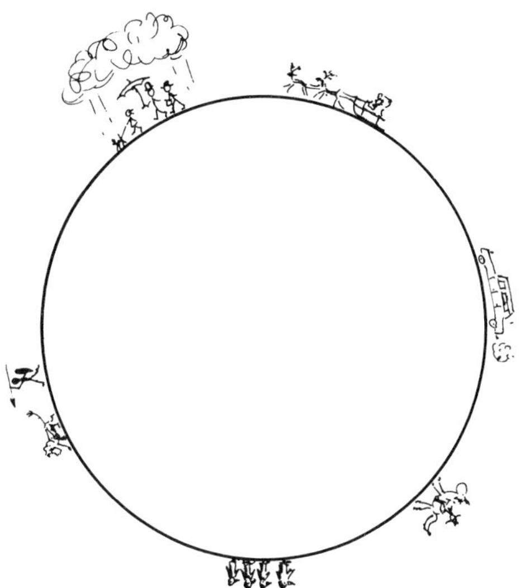

Draw the direction of gravity that allows all these people to remain on our planet earth.

The force of gravity acts on all countries of the world and is directed towards the centre of the earth.

The earth is surrounded by an atmosphere of air which we breathe. We need to breathe air to say alive, but why does the air remain around the earth and why is there no air out in space?

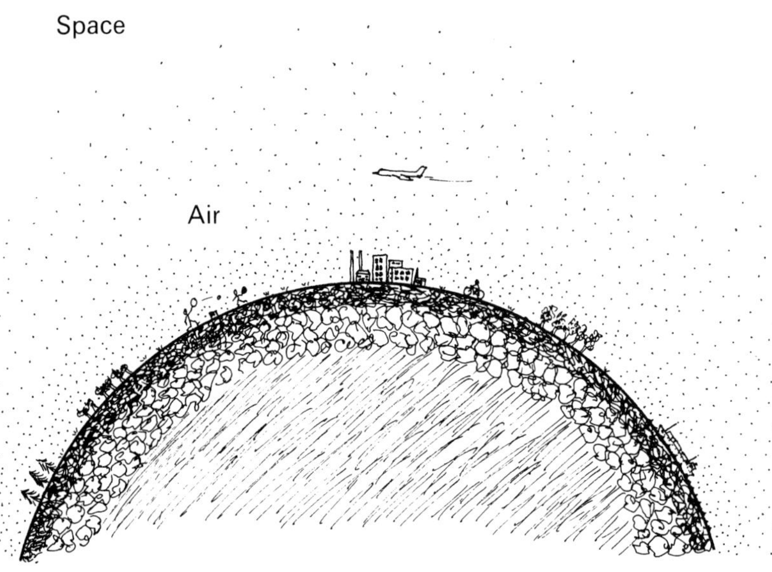

Again it is the force of gravity which holds all the air round the earth and stops it from floating into space.

Draw some lines on the diagram to show that this is so.

How Can We Escape Gravity?

Take a ball and throw it up a few metres.
The ball travels up, stops, and then falls back to earth.
If you went outside and threw the ball as high as you could it would still come back.
You do not have enough energy to throw the ball out into space.

To escape from the gravity of the earth we would need a rocket.

It takes a lot of energy to escape from the earth's gravity.
But as we move away from the planet the force of gravity gets less and less.

If we were out in space, away from the earth and other planets, then we would feel no pull of gravity. We would be free to float about. Have you seen the TV pictures of astronauts floating in space stations?

When this happens we say that they are *weightless*. If there is no force of gravity pulling on an object then it will weigh nothing. Although it weighs nothing we can still see it and touch it, it is still there.

We can still see the weightless astronaut in his rocket. He might weigh nothing but he still has *mass*. We all have mass. Mass is the measure of the amount of matter inside something. All objects contain matter and have mass.

If you were floating in space you would still have the same mass as you have on earth, but you would have no weight as there is no gravity to pull you down.

Take the ball of plasticine and the ball-bearing. Hold one in your right hand and one in your left hand. The ball-bearing feels heavier although it is the same size as the ball of plasticine. Why is this?

Ball-bearing

Plasticine ball

The ball-bearing contains more matter than the plasticine ball, so the ball-bearing has more mass.

This means that on earth the ball-bearing will weigh more than the plasticine ball (you can feel that it weighs more) because it has more mass. In space they would both weigh nothing but the ball-bearing would still have more mass.

On earth the man has mass and weight, because his weight is a measure of the force of gravity on him.

In space the same man is weightless (weighs nothing) because there is no gravity pulling on him, but he still has the same mass.

52

Fill in the blank spaces:

It is the force of _________ which pulls things towards the earth.

All round the planet gravity is directed towards the _________ ____ _________ _________

To escape from the earth's gravity we need to use a lot of _________

In space, without gravity we would be free to float because there would be no _________

On earth and in space our _________ remains the same.

To think about
Can we make use of this gravity in any way to help us?
Are there any ways of harnessing gravity to do useful work?
Could the falling power of waterfalls be used for anything?
Do swinging things use gravity?

PENDULUMS
(The Use of Gravity)

Information for Teachers and Parents

Skills and Processes Included in this Section
Measurement of time
Measurement of length
Some maths skills
Counting

Words and Vocabulary Needing Explanation
Swinging
Period of swing
Pendulum

Materials Needed
About three metres of string or cotton
Something to use as weights, e.g. coins or plasticine
Sellotape
A few empty matchboxes
Paper clips
Watch or clock with a second hand, digital watch will do
Empty washing-up liquid bottle
A cupful of fine dry sand or dry powdered salt
Sheet of newspaper or card

Pendulums, 1, 2, 3

In the year 1583 a man called Galileo Galilei was attending a service at the cathedral in Pisa. Hanging from the roof of the cathedral were many lamps suspended by long chains. When one of the lamps was set swinging Galileo sat and watched it. He noticed that the lamp took the same amount of time to complete each swing. As there were no watches 400 years ago he used his pulse to time the swings. When he left the cathedral, Galileo went home and started to experiment with pendulums.

In this section you will find out more about pendulums and how their motion affects us all.

You will need: Paper clip
                  String
                  Matchbox containing a 10p coin

What to do:

Take a length of string, just shorter than the height of a door. Open up a paper clip to make a hook, S, and tie this to one end of the string. Take a matchbox and attach this to the other end of the string. Now hook the paper clip over the top of an open door frame; this is to be our simple pendulum. (You may need to tape the clip to the door)

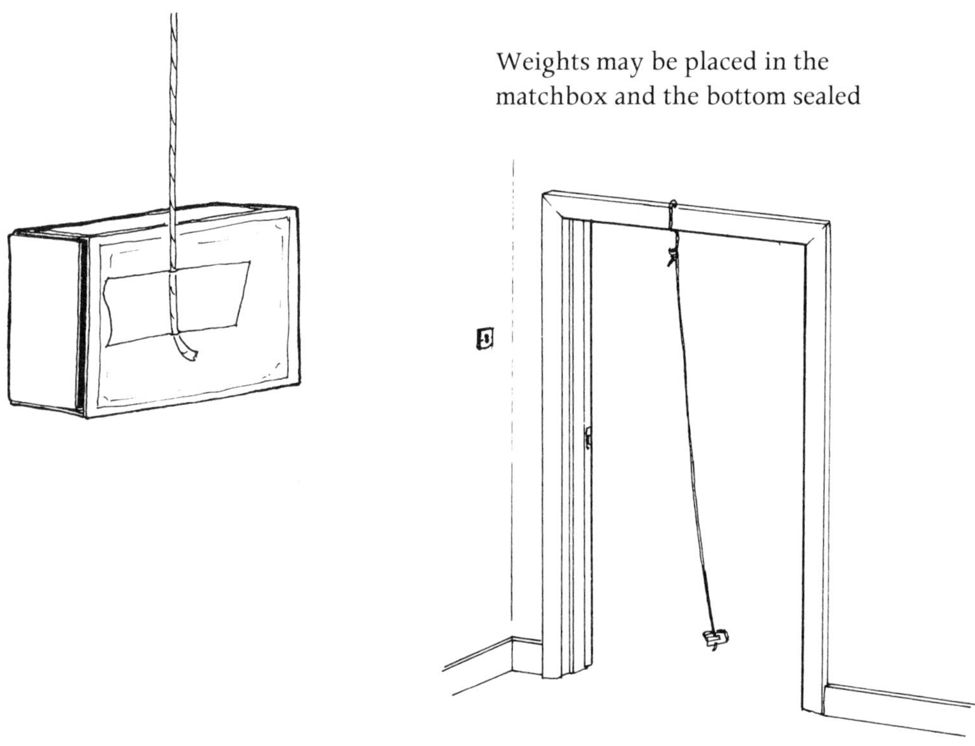

Set up the pendulum and place a weight of one 10p coin in the matchbox. Pull the matchbox gently about 10 cms to one side and let it go. The moment you let it go, start timing how long it takes to make ten swings.

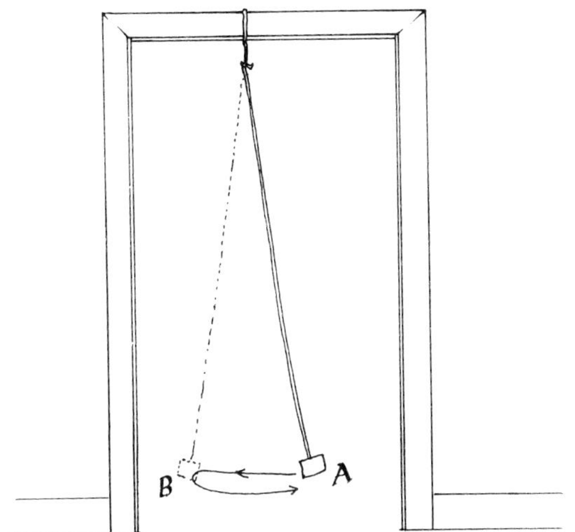

One whole swing is the movement from point A to point B and back to point A again.

Repeat the experiment by pulling the matchbox 5 cms to one side and timing 10 swings and then 20 cms to one side and timing ten swings. Write your results in the table below.

Once you have recorded the time for ten swings you can calculate the time for one swing (by dividing the time for ten swings by 10).

Distance matchbox pulled to side	Time for 10 swings	Time for 1 swing
5 cms		
10 cms		
20 cms		

EXPERIMENT 2

Set the pendulum up again, keeping the same weight in the matchbox. This time we are going to vary the length of the pendulum. By wrapping more of the string around the hook we can vary the length between one full length and a third of this length. Take the time for ten swings at:

One
third
length
Two
thirds
length
One
full
length

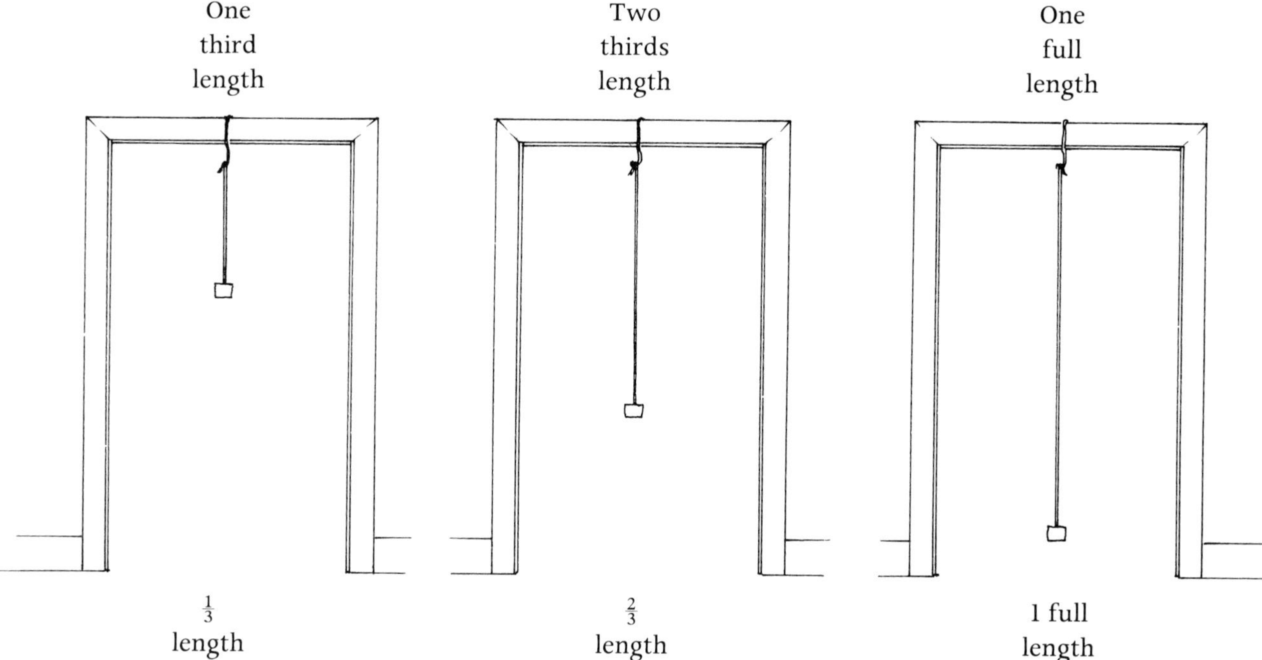

$\frac{1}{3}$
length
$\frac{2}{3}$
length
1 full
length

Write your results in the table.

Length of string	Time for 10 swings	Time for 1 swing
$\frac{1}{3}$ length		
$\frac{2}{3}$ length		
1 length		

EXPERIMENT 3

Put the pendulum back to its *full* length. This time we are going to alter the weight in the matchbox. Take the time for ten swings with one weight of a 10p coin, then with two weights, two 10p coins and then three weights, three 10p coins. Write your results in the table below.

Number of weights	Time for 10 swings	Time for 1 swing
1		
2		
3		

In each of the experiments you have worked out the time for one complete swing: the *PERIOD*.

Look back at the experiments and the tables you have filled in. For Experiment 1, do you think the period changed when the distance the matchbox was pulled to one side was changed?

In Experiment 2, did the period change when the length of the pendulum was changed?

In Experiment 3 did the period change when the weight in the matchbox was changed?

Things to think about

What makes the pendulum swing backwards and forwards? Why does it eventually stop and come to rest in the centre of its swinging position?

Would the pendulum experiment work in a spacecraft under weightless conditions?

If we took our experiments to the moon where gravity is much less than on earth would the pendulums still swing at the same rate?

Where, during your outings to the shops and park, do you see a use for swinging things? Make a list of these.

A DOUBLE PENDULUM

Make two equal length pendulums and wind each one around a pencil or piece of dowel.

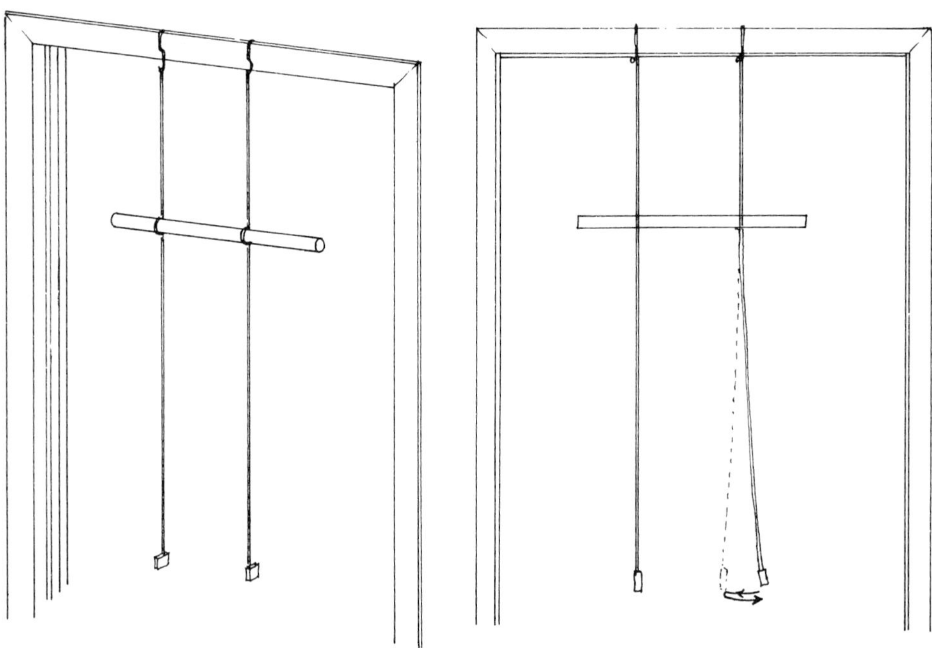

Set *one* of the pendulums swinging and leave it for some time.

Notice what happens to each of the pendulums in turn. What do you think would happen if a third pendulum were added? Try this and see. What if the rod were put at a lower position? (Remember Experiment 2.) Move the pencil down and see what happens.

Your Findings

You should have found that the period depends on the length of the pendulum. Actually the period depends on the square root of the length.

We can write (approximately): PERIOD $= 2 \times \sqrt{L}$

Where L = length of the pendulum and is measured in metres.

Set the experiment up again and choose a length of one metre (square root = 1), and test to see if the above formula is correct (what are you expecting the answer to be?).

What use can we make of pendulums which swing at known periods?
Can they be used for accurate timing?

PATTERNS WITH PENDULUMS

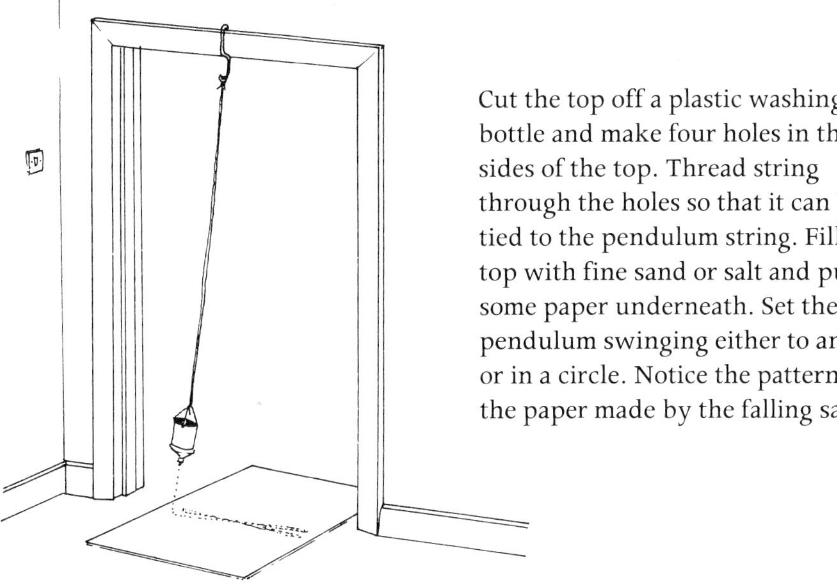

Cut the top off a plastic washing-up bottle and make four holes in the sides of the top. Thread string through the holes so that it can be tied to the pendulum string. Fill the top with fine sand or salt and put some paper underneath. Set the pendulum swinging either to and fro or in a circle. Notice the pattern on the paper made by the falling sand.

Try making different patterns by altering the length of the string or by fixing the pendulum like this.

EXPERIMENTS TO DO IN THE PARK

When you are using the swings in the park you can make yourself go faster by swinging you legs in rhythm with the swing.

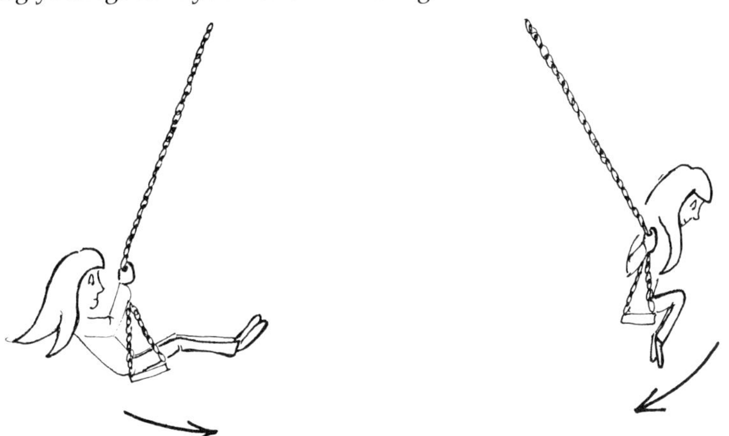

Why does this make the swing go faster?
What have you got to do to slow down?

Now find a friend who is a different weight from you. Get someone else to give you both an equal push. If neither of you swing your legs, who would you expect to swing faster? (Remember Experiment 3.)

COLOUR AND CHEMISTRY

Background Information for Teachers and Parents

Skills and Processes Included in this Section
Manipulative skills
Measuring
Constructing
Writing
Explaining and predicting
Observing and inferring
Experimenting and testing
Writing and drawing

Words and Vocabulary Needing Explanation
Chromatography
Dissolving and separation
Prism
Absorbent and soaking
Opaque and transparent
Refraction

Materials Needed
Various coloured paints
Felt tip pens
Absorbent paper (blotting paper or chromatography paper)
Writing paper and pencil
Glass prism
Shoe box
Cotton wool
Small containers for water
Smarties or other coloured sweets

For the Teacher
The purpose of this section is to give all the details needed to carry out simple colourful and interesting experiments with groups of pupils who do not normally undertake chemical experiments. Such pupils may be the slow learners, or pupils with learning difficulties, and pupils with special educational needs.

The details are written in a manner to allow the pupils to record their observations.

The wide range of experiments should enable pupils from a wide range of backgrounds to find some experiments suitable for everyone. Some pupils will use this section as no more than a 'colouring in' book of their experimental findings; others will be able to carry out a more open-ended set of experiments and this can give them some sense of direction.

There are information notes below for the teachers on those experiments which were thought to warrant them.

When these experiments were tested with a wide range of pupils, some pupils coloured in the title pages and some other diagrams and this was the limit of their ability and motivation; other pupils rushed to proceed beyond the designated areas and these pupils eagerly gave illustrations and written work beyond their normal expected work load.

One class teacher commented, after trying these experiments with a class of physically handicapped pupils of secondary age: 'It proved to be the most successful project I have attempted in 14 years of teaching. Pupils were well motivated throughout. At a parents' evening the parents were very keen to find out what CHROMATOGRAPHY was, as the pupils came home talking excitedly about their experimental work.'

The detailed chemistry and physics of the individual experiments are not the main purpose, but the science is used as a vehicle for experimentation, achievement and success by the pupils.

BACKGROUND INFORMATION ABOUT THE EXPERIMENTS

For Teachers
Sunlight (or white light as it is called) is made up of a series of colours and when these are combined together they make up white light.

When white light passes through a glass prism, a raindrop, or other transparent object, the light is bent and each colour is bent (or refracted) to a different extent, so being separated into its component colours—violet light is bent the most and red the least. The colours always appear in the same order.

In the experiment using prisms, use the triangular equilateral prism.

Mystery Box
You can buy the different coloured transparent papers (or filters) from a supplier, e.g. Arnolds, to cover the windows if you desire, rather than using sweet papers.

The object of the experiment is to get the pupils to predict the colour of the object in the box, rather than fully explain why.

Mixing Colours
The multi-coloured disc is called the Newton's disc and made by approximately seven equal divisions, each coloured as shown in the experiment. This can be made by dividing the circle using a protractor: $\frac{360}{7} = 51 \cdot 5°$

Butterflies, Fishes and Detectives
These experiments use safe chemical dyes used in foods and felt tip pens. A good collection of water-soluble pens and a box of Smarties, or a collection of food dyes, is needed.

Some teachers have found that pupils (and even the reluctant ones) wanted to write about these experiments, so opportunity is given to allow them to do so.

Chromatography Experiments
Present the chromatography experiments to the pupils in as dramatic a way as possible, eg:

> Sssh! Is anyone watching (looking over your shoulder and outside the door)? I'm a spy working for 'Smiths' shops and I want to know what colours are used in the felt tips made by 'Boots'. How can I do it? (Ask the pupils for suggestions.)
>
> I've read in a book that if you put a drop of ink on a piece of absorbent blotting type paper and then let water soak across it, some inks split up into different colours. Like this:

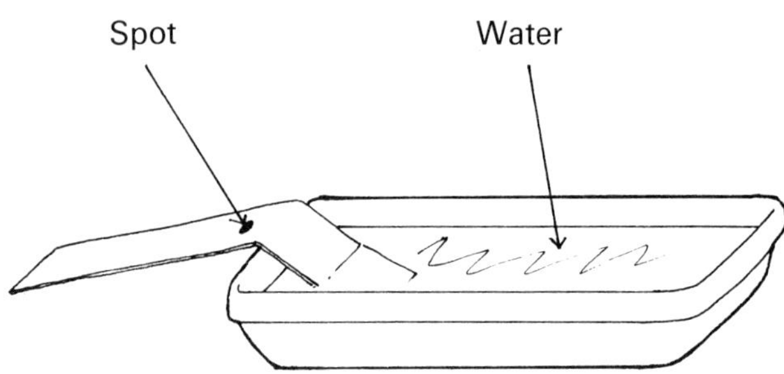

Give the children a series of black pens of different makes. Ask them, do all the black pens use the same inks?

You can do the same for any other coloured pen also (but try them yourself first as only water-soluble inks separate out by this method of chromatography).

Don't let the ink spots dip under the level of the water as the dye dissolves in the water. You want the water to soak along the paper and push out the separate colours from the dye mixtures.

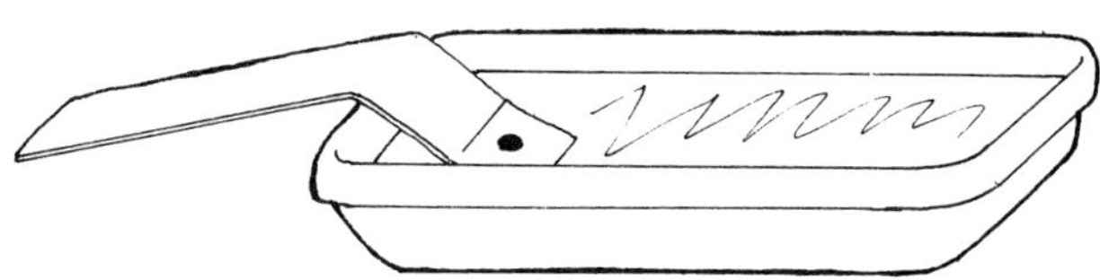

Spot in wrong position
Not under water level

Another spy story can be woven around the Smarties experiment. Let the pupils write their own spy story at the end of these experiments. Try this method for the Smarties experiment:

Take a circle of filter paper and lick the surface of the sweets and put spots in a small circle like this:

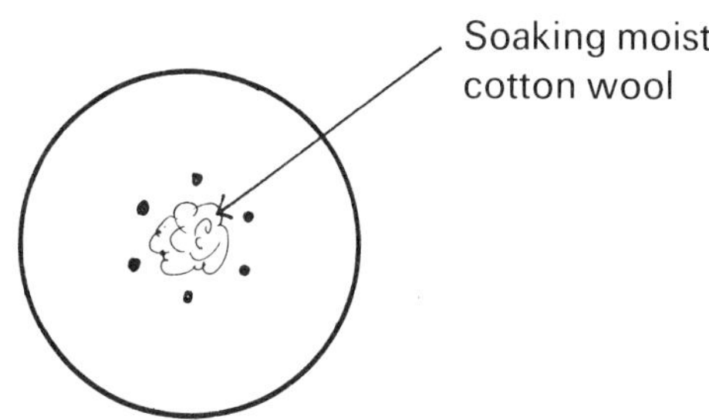

The soaking water from the wet cotton wool pushes across the paper and separates the colours all at once.

Use felt tip pens or sweet dyes. Paints do *not* usually contain mixtures of pigments, so don't use paints for chromatography experiments.

Playing Detectives
Make your own forged cheques (as authentic as possible) on absorbent blotting or filter papers. You can use two different inks to put extra 0s on the figures and in writing in the amount, e.g:

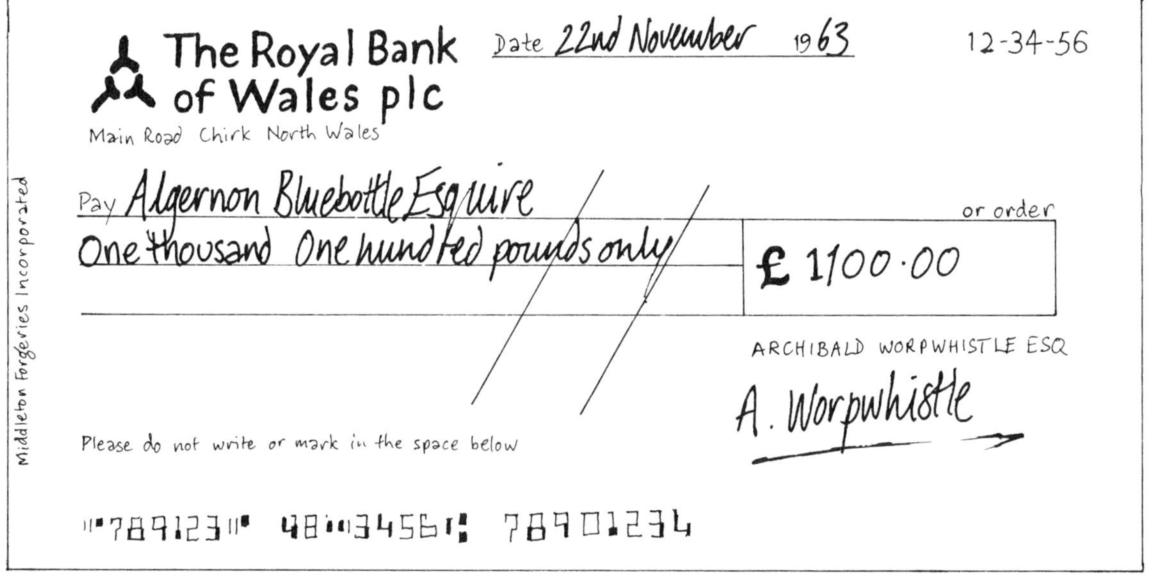

The pupils might put a piece of wet cotton wool under the writing or amount, or they might try to blot the inks into another piece of paper for separation purposes.

This experiment usually leads to interesting conversations.

Butterflies and Fishes
These experiments are designed for pupils to use and merge science with art.

The colours separate to give some excellent results. Dry the paper with a hair-dryer or on a radiator.

Paper
We have used absorbent paper for these experiments; this is necessary as most other papers have been coated with china clay to prevent the inks spreading.

COLOUR

A
PROJECT BY

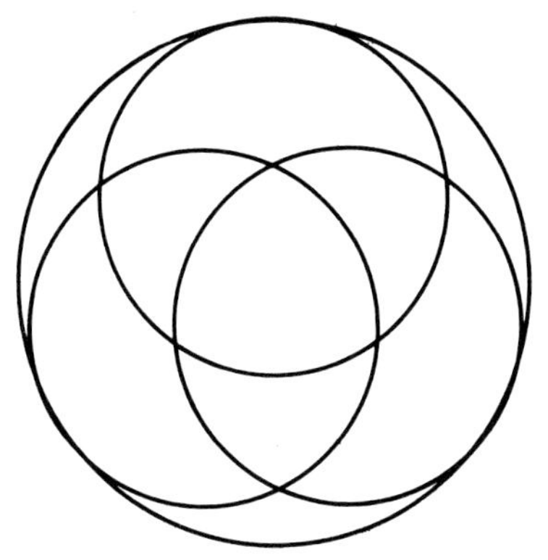

Using Your Eyes

Give each pupil a pallet of coloured paints and a few paint brushes.
Ask each pupil to paint in their colours on the painting diagram shown below:

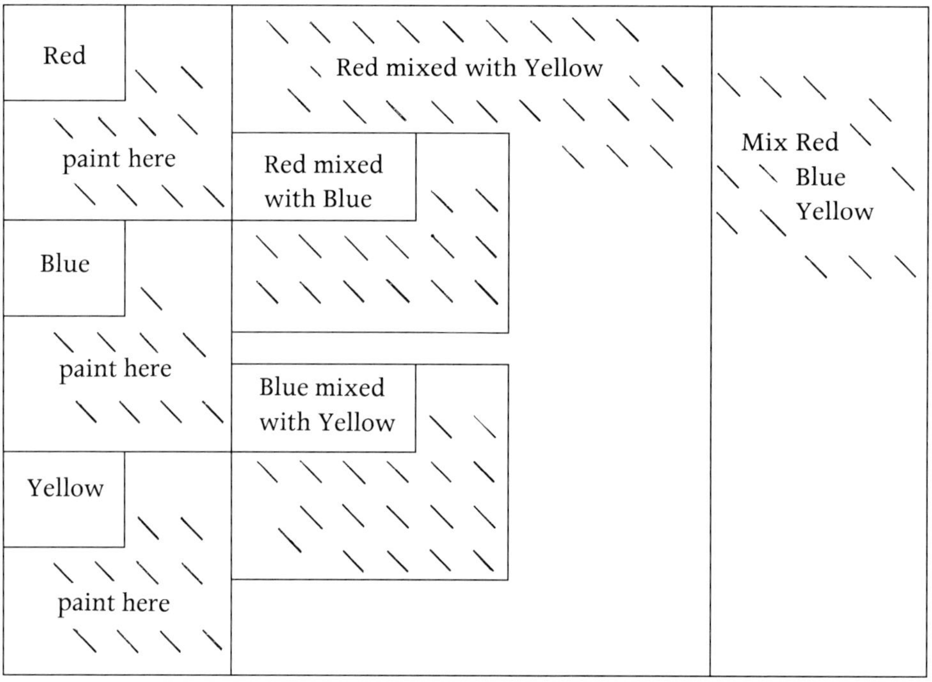

<table>
<tr><td>Red
paint here</td><td colspan="2">Red mixed with Yellow</td><td rowspan="3">Mix Red
Blue
Yellow</td></tr>
<tr><td>Blue
paint here</td><td>Red mixed with Blue</td><td></td></tr>
<tr><td>Yellow
paint here</td><td>Blue mixed with Yellow</td><td></td></tr>
</table>

Which colours when mixed give green? _____________ and _____________
Which colours give orange? _____________ and _____________

Do your own shapes and designs below

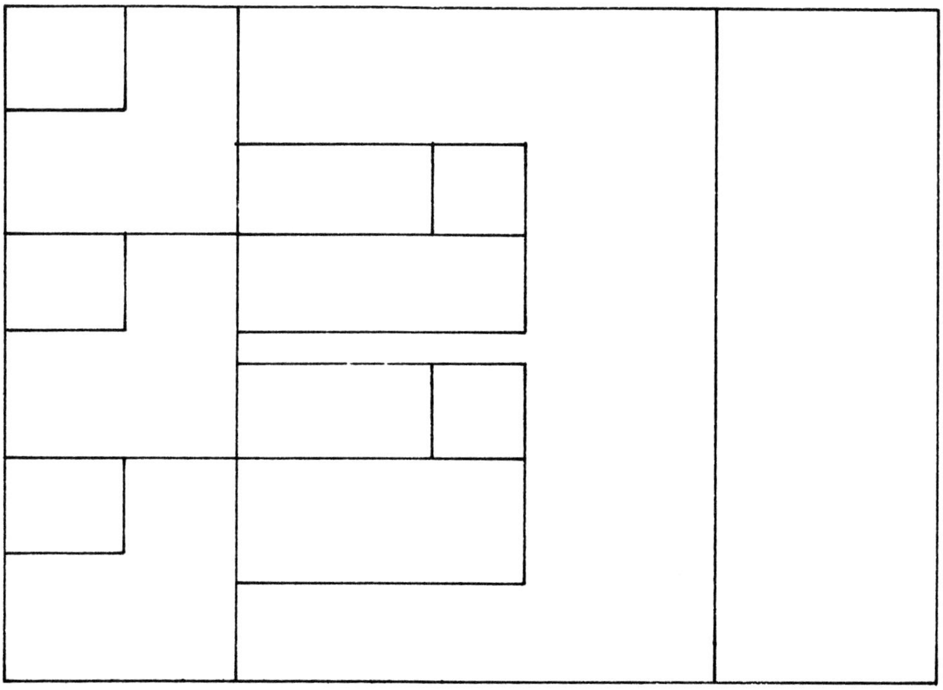

Look around the room and outside. How many colours can you see?
Fill in the boxes below with the names of things of that colour.

RED	BLUE	YELLOW

GREEN	ORANGE	BROWN

PURPLE	WHITE	BLACK

Colours of the Rainbow

Everyone has seen a rainbow, but can you remember the colours, and the order they are in?

1   Some people remember the order by using this sentence:
Richard Of York Gave Battle In Vain
R = Red, O = Orange, Y = Yellow, G = Green, B = Blue, I = Indigo, V = Violet

2   Colour this in the correct order:

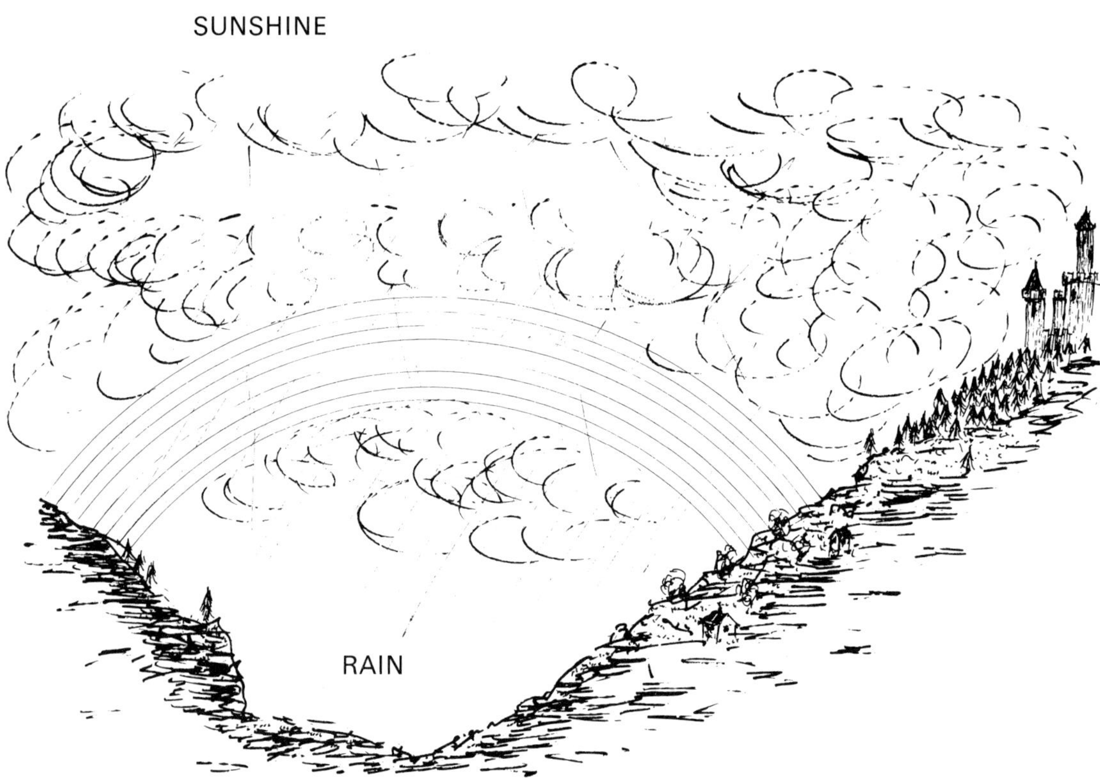

3   Could you make up a sentence to help you remember the colours of the rainbow? They must be in order.

Some pupils' sentences were:

    Rats On Your Grave Bite Infected Vampires.

    Robins On Your Grass Bring In Votes.

Here are a few words you could use:

    V = vodka; voice; volume; votes; vacuum.

    G = gas; glass; ground; grumble; ghost; go; got.

    R = rot; Ronald Reagan; rug; rust.

You write some words starting with:

O =

Y =

B =

I =

MAKING THE COLOURS OF THE RAINBOW

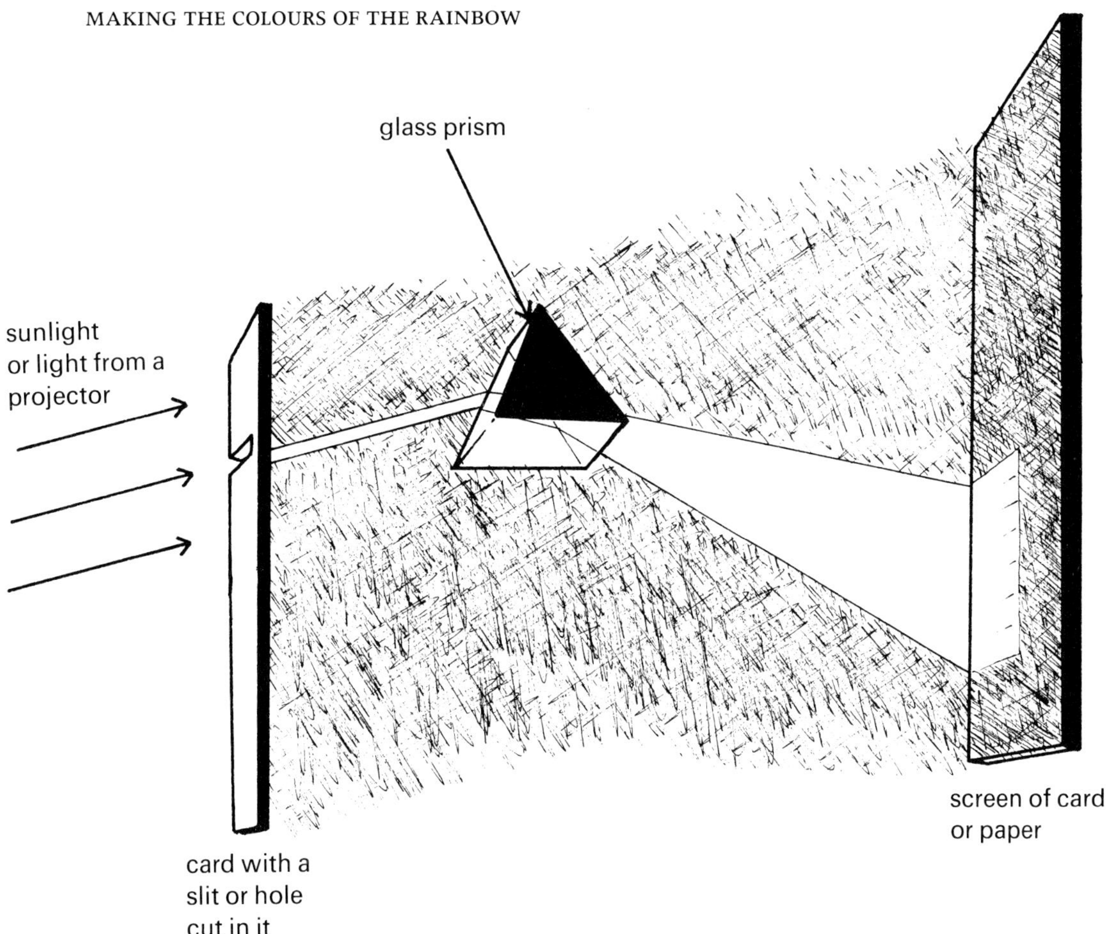

Colour in the colours you see.

Can you explain how the rainbow's colours are formed when it rains and the sun is still shining?

(Hint . . . the raindrops act like prisms.)

Information

Light is made up of seven colours. (You should know by now what they are.)

When light goes through *transparent* things it gets bent. Some things bend light a little, some things bend light a lot.

Some things bend light so much that it splits up into its seven colours.

Another name for this bending is *refraction*.

Using a dictionary, find the meaning of:
1   Transparent—
2   Opaque—

Fill in the table below

SOME TRANSPARENT THINGS	SOME OPAQUE THINGS

The Mystery Box

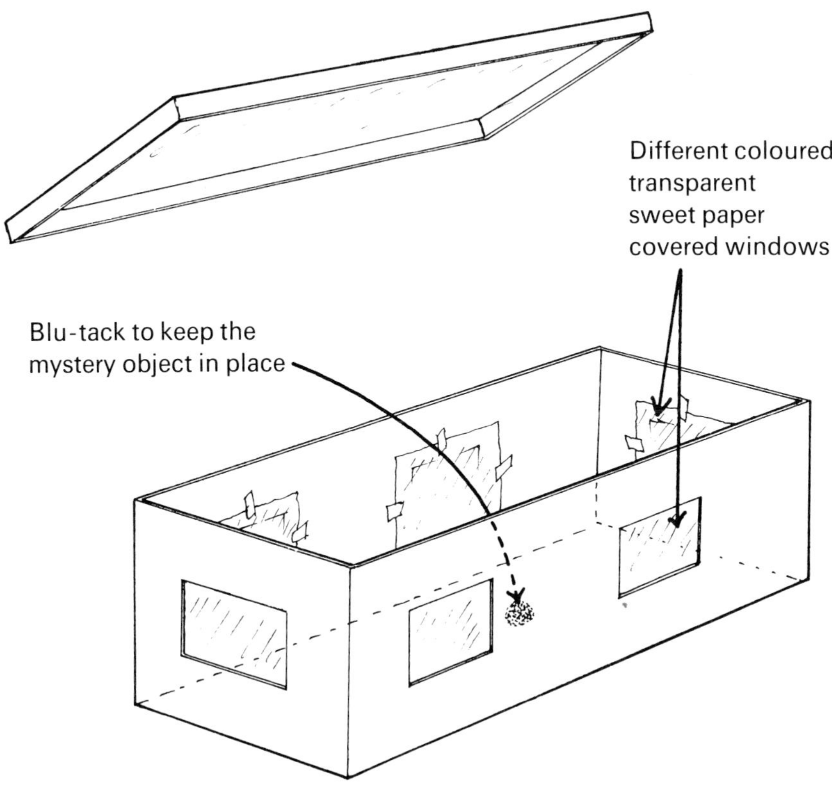

Take a shoe box (or any other box) and cut out small windows. Cover each window with a different coloured transparent paper (e.g. from sweets, etc.).

Put inside a coloured object like a pencil or coloured sweets.
Put the lid on. Give the box to your friends and ask them to guess the colour of the object. They will look through the different coloured windows and see different colours.

Is there any pattern in the colours they see through the different windows?

To think about
Do coloured clothes look the same in shop lighting and outside in the daylight?
Do cars look the same colour in daylight and street lights at night?
Do clothes look the same colours under the different lights in a disco?
What happens to the colour of people's clothes and faces when you turn the 'colour' control knob on the TV?
How do you know where to turn it to get the correct colour of the picture?

Spinners and Mixing Colours

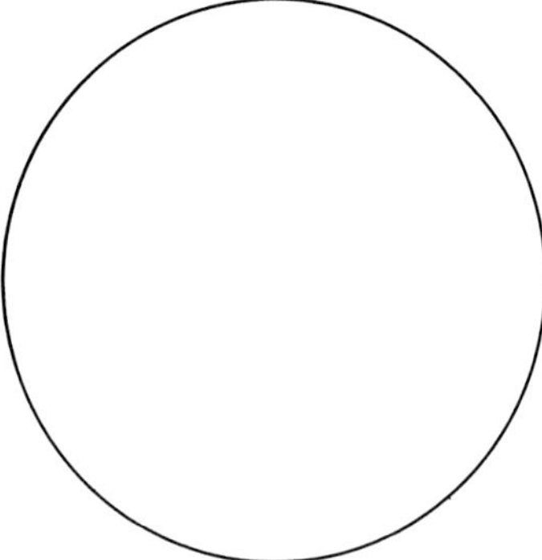

Cut out several circles,
about this size, from
white card.

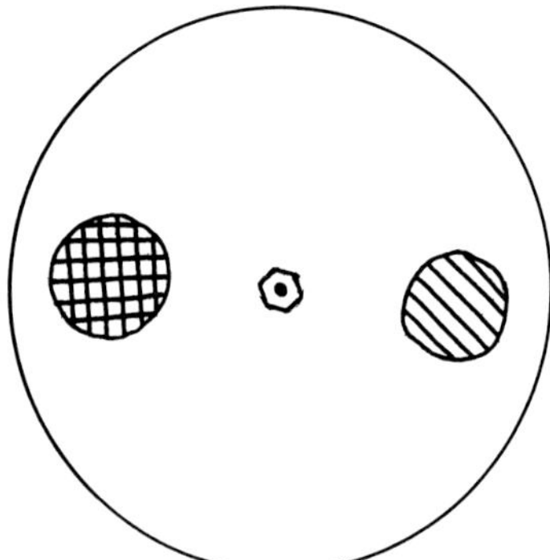

Put on two blobs of
different colours of paint
or felt tip pens

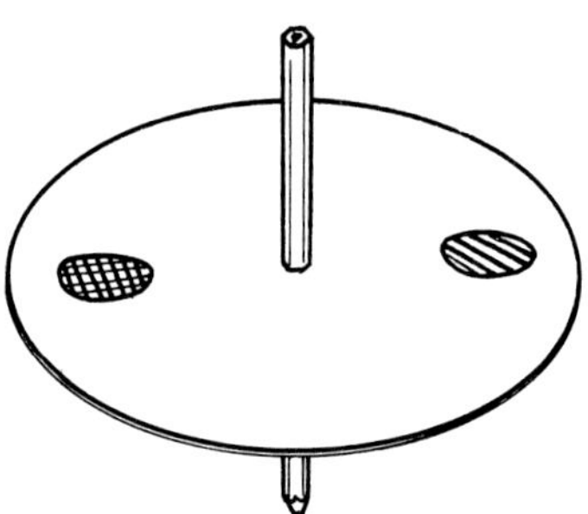

Push a sharp match or
pencil through the centre
of the circle.
THEN SPIN

If you have done it correctly, you should see a change of colour. Record your results on the next page.

If you are feeling very brainy, you could make an electric colour-spinner by copying the diagram below.

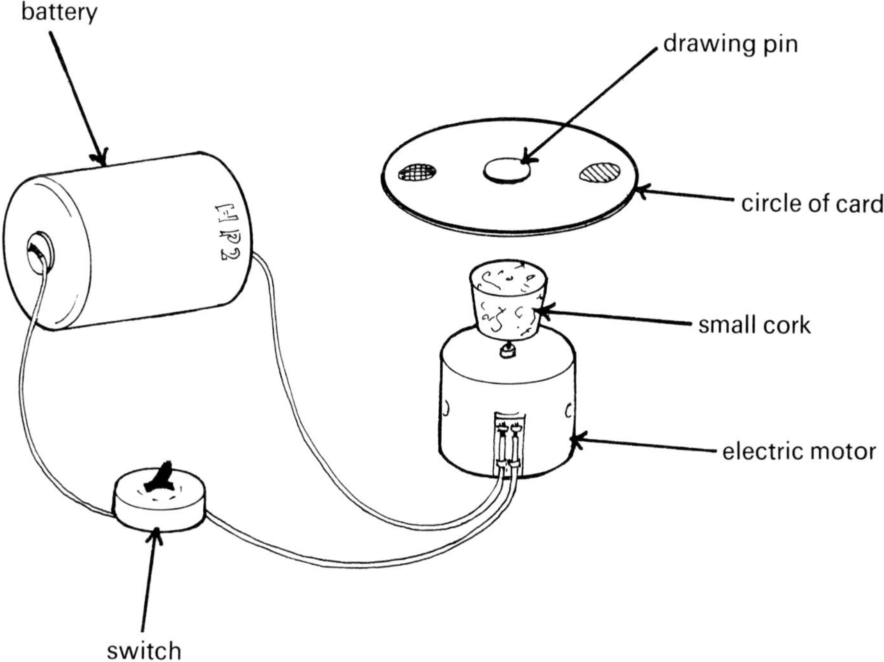

Mixing Colours

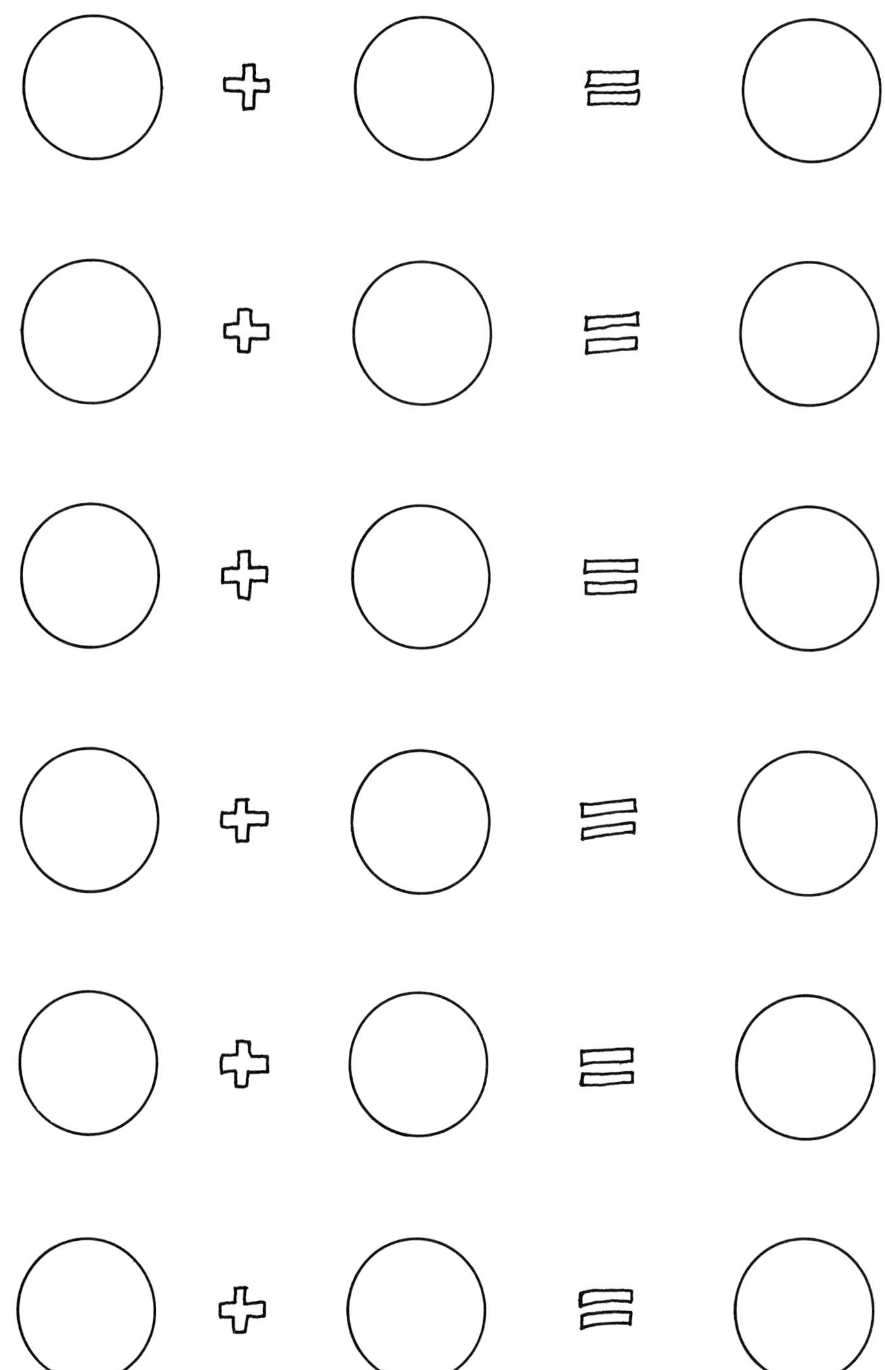

Now try this

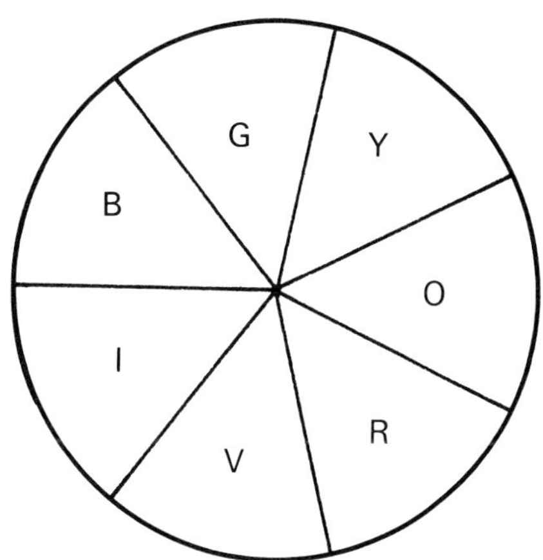

Cut out a circle of white card. Divide this up into seven approximately equal sections (the teacher will tell you how to do this). Colour these in using the colours of the rainbow.

Y = Yellow
O = Orange
R = Red
V = Violet
I = Indigo
B = Blue
G = Green

Spin this disc either on a pencil or using the small motor.

What happens to the colours on spinning the disc?
What would happen if you missed out one or more of the coloured sections?
Predict your answer here ________________________________ Now try to see if you were right.
Try leaving out others also.

To think about
Instead of using colours, use other patterns on the spinning disc, for example:

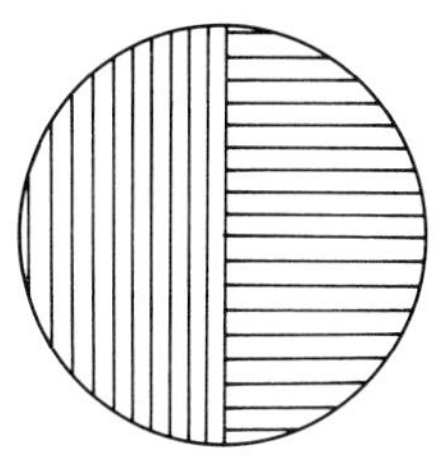

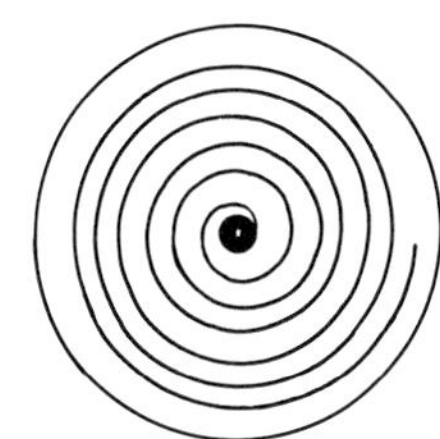

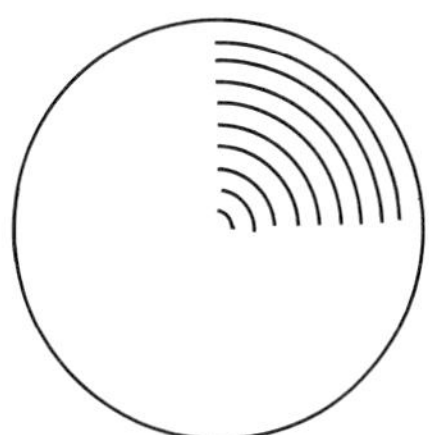

You design some and try.
You design one. No one else has ever tried your special design, so you are discovering things today that no one else has seen. You had better write down what you see, in your book.

Chromatography

Here is an experiment for you to do yourself.

WHAT COLOURS ARE IN FELT TIP PENS?

What I wanted to find out

What equipment I used

What I did

What I saw

WHAT COLOURS ARE IN SMARTIES?

What I wanted to find out

What equipment I used

What I did

What I saw

Now do this

1   List the colours of the dyes in the sweets.

Colour of Smartie	Colours of dyes used
Orange	
Yellow	
Light Brown	
Dark Brown	
Purple	
Green	
Red	

2   Supposing you wanted to make a *chocolate* coloured sweet, which food dyes would you mix?

3   If you wanted to find out what food dyes were used in making raspberry or strawberry jam, how would you set about it?

Playing Detectives

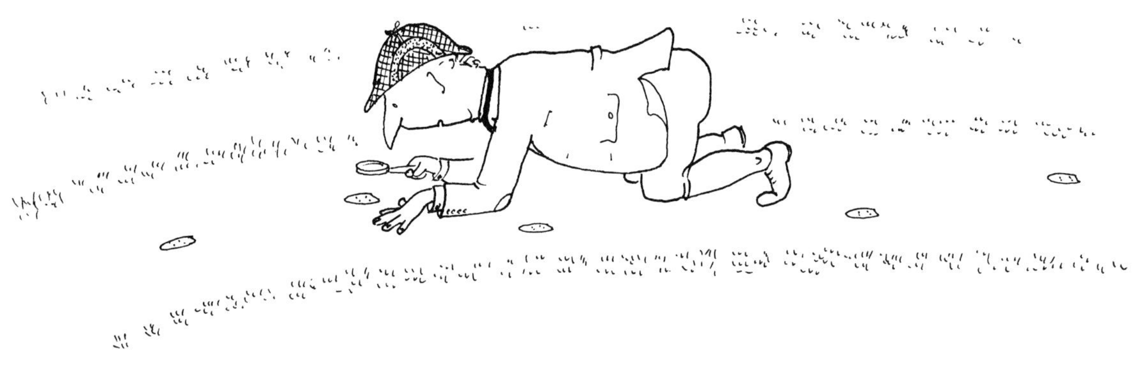

USING CHROMATOGRAPHY TO FIND THE FORGED CHEQUE

The problem!
You have two cheques. One is forged and one is not. Can you find out which one is which?

What I did

What I saw

My decision

Colourful Fish

This is what you do
Cut the shape below out of white filter paper, or blotting paper.

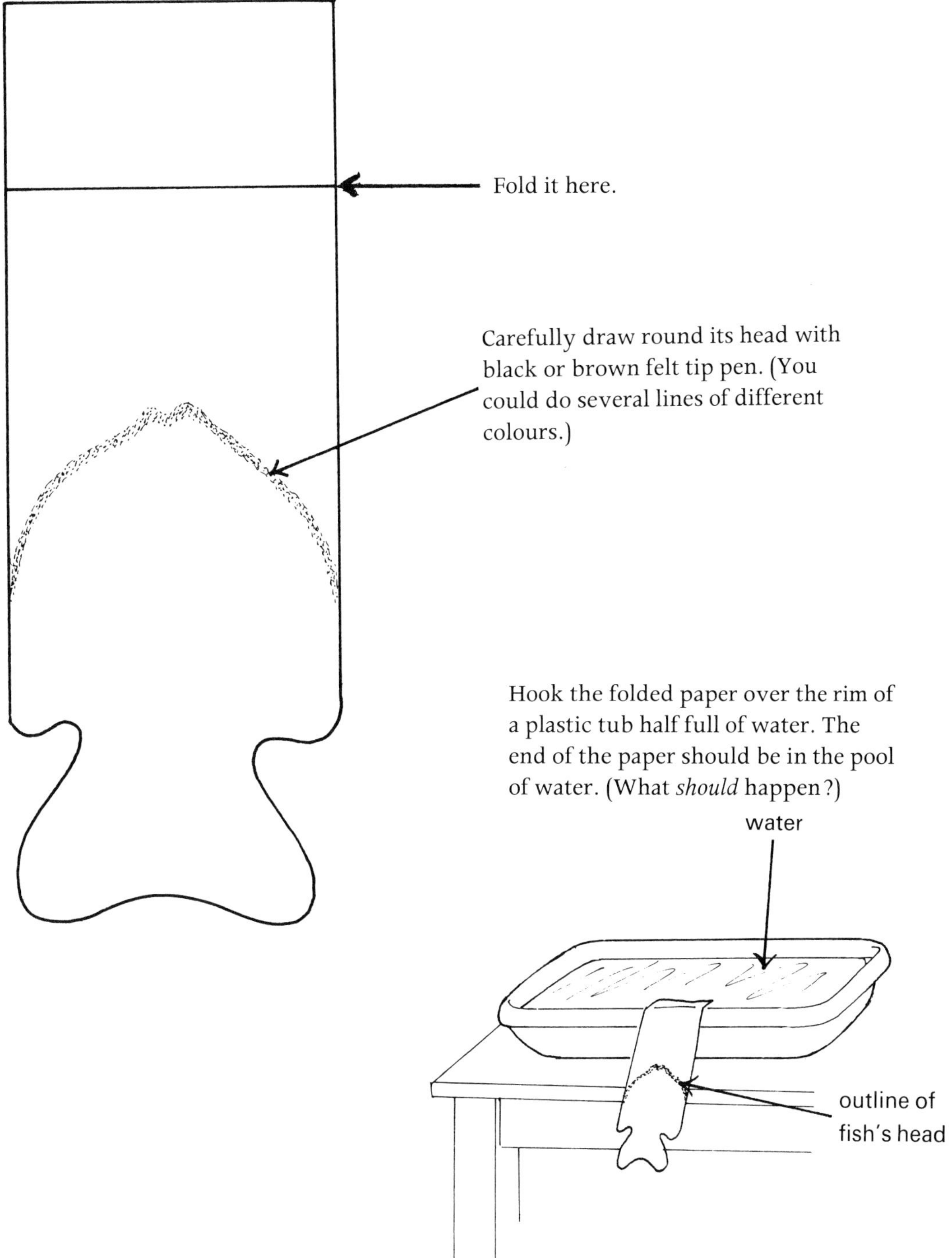

Fold it here.

Carefully draw round its head with black or brown felt tip pen. (You could do several lines of different colours.)

Hook the folded paper over the rim of a plastic tub half full of water. The end of the paper should be in the pool of water. (What *should* happen?)

Repeat the process using different coloured pens. Dry the paper when the colours are what you want.

Colourful Butterflies

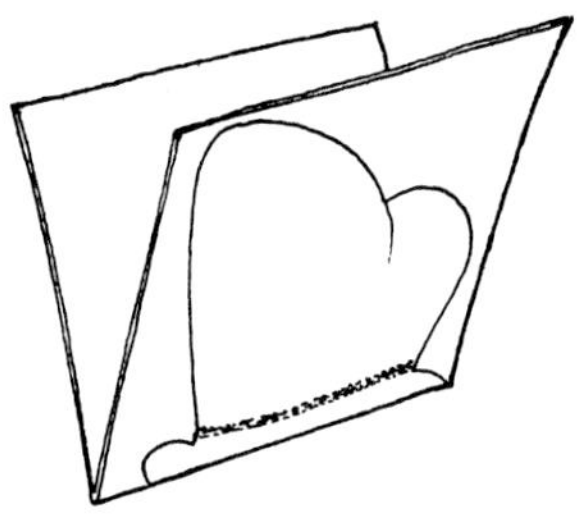

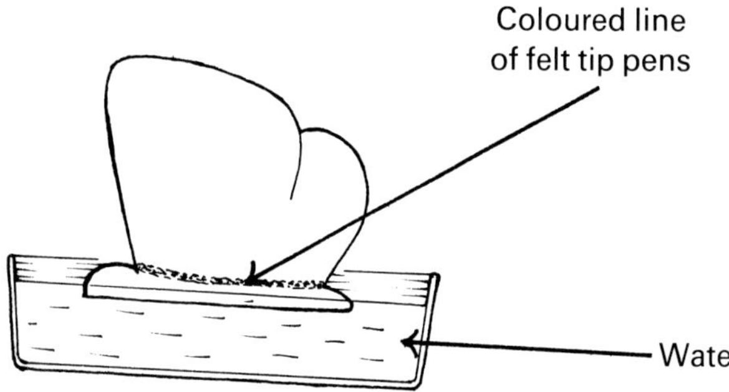

You try drawing some butterfly shapes or birds and see how they look when using Chromatography.

Here's a hint: Fold your paper in half and cut out the shape you want.

Colour around where the body joins the wings.

Dip the body in the water but don't allow the ink to touch the water.

Allow the water to soak up the wings until you have the colour you want.

Then dry it on the radiator or with a hair dryer.

Alternatively, you could lay the butterfly on the table and put wet cotton wool along the centre fold line.

Try other shapes—animals, birds, insects, flowers.

Display. You can display your fish or bird as a mobile.

Arty Crafty

This activity involves making colourful backgrounds onto which, when dry, you can stick a suitable cut-out shape, e.g. a rocket shape, or the space shuttle landing.

You will need: A sheet of filter paper, or blotting paper (about the size of this page)
Felt tip pens
A thin cane or rod
An aquarium or washing-up bowl
A bull-dog clip

What to do:

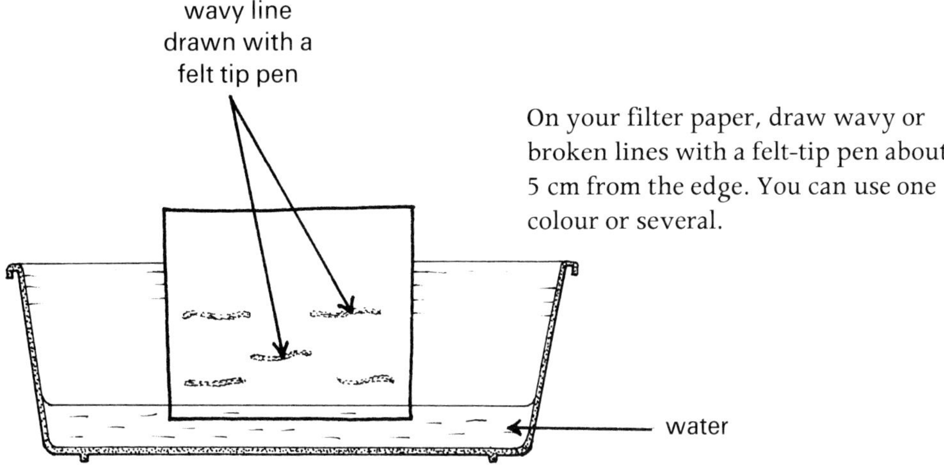

On your filter paper, draw wavy or broken lines with a felt-tip pen about 5 cm from the edge. You can use one colour or several.

Put 3–4 cm of water into the bowl and suspend the filter paper as shown above. The water level MUST NOT touch the ink lines, it must only soak across them.

Watch the colours move upward and remove the paper for drying when they are near the top.

When dry cut out black shapes of, say, a rocket and place it on the background you have made. Try other shapes and other backgrounds.

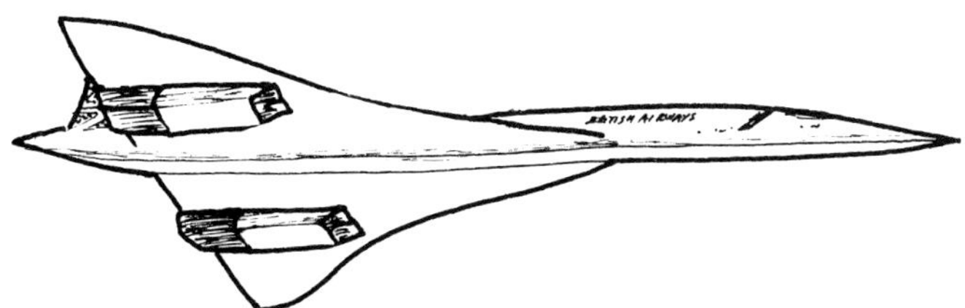

ART WITH MOVING SHAPES

Try these different pictures. Draw the shapes with felt tip pens.

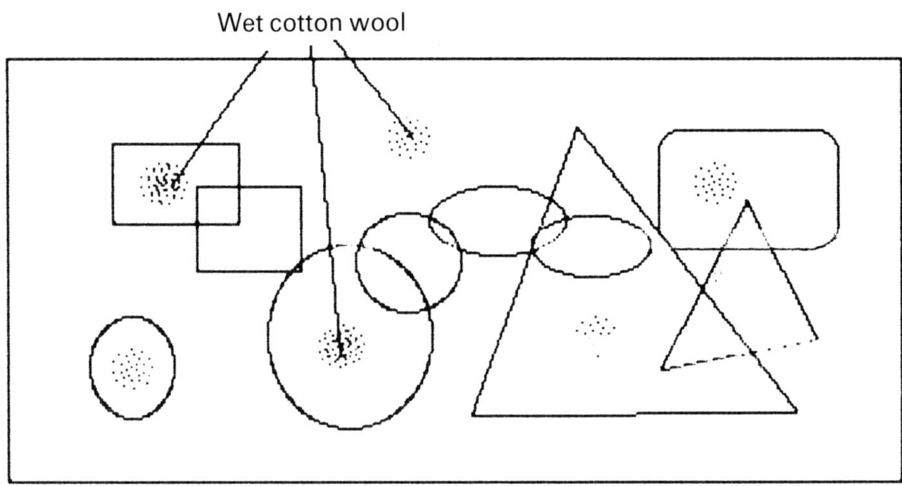

Put wet cotton wool anywhere on your picture.

Draw with felt tip pens

At a Snail's Pace

Materials: Absorbent paper (e.g., either blotting paper, filter paper or white strip of unprinted newspaper)
Felt tip pens (water soluble)
Saucer of water

What to do:
Cut a strip of absorbent paper about 20–30 cms long and draw a pencil line about 5 cms from one end.

Draw a Snail behind this line facing towards the long end.

Then draw a pencil line 1 cm from the other end. We will call this the finishing line.

Dip the end behind the Snail in the saucer of water so that the water soaks along the paper.

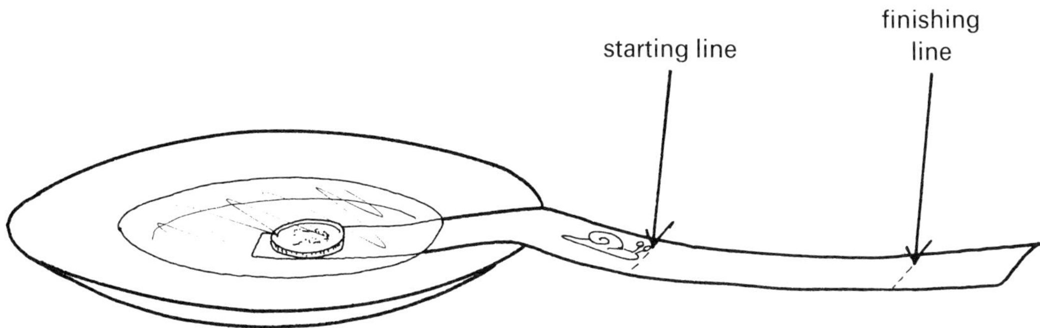

Watch how long it takes for the Snail to leave a trail to cross the finishing line.

Extension

Suppose you draw two or three Snails, each in a different colour. You could soak water along the paper track and see which one was the winner.

Predict which one you think will win! Or perhaps they will arrive at the same time.

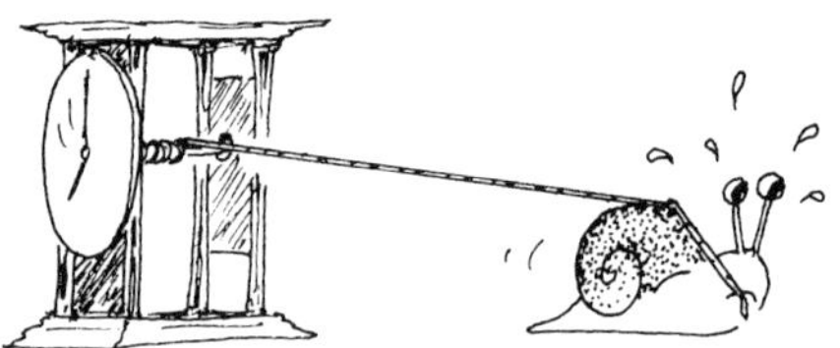

water soaking snail's pace clock

Comments by David A. Coleman

The term was not long enough to complete all lines of enquiry suggested by either the pupils or myself.

Given more time I intended to experiment with dyes. I have done Tie'n'dye and Wax Resist dyeing in the past—both these would have been useful support activities for the Project.

Animal and insect camouflage was another area we touched on only briefly—this too deserved a deeper examination.

Colour as a warning (traffic lights, fluorescent arm bands, lamps on Police cars, indicator lights, etc.) was another area which could be explored.

Skin colour also has interesting possibilities.

I should also have liked to include a visit to a factory producing paints or dyes—there are several in our area.

Even though many of my pupils suffer severe physical disabilities, few, if any, of the experiments or activities were beyond their limitations. No heavy apparatus was needed and all experiments were table-top based.

A few who, because of intellectual or physical problems, found the keeping of notes difficult, were well enough motivated to discuss their experiments both with myself and with the adult helpers.

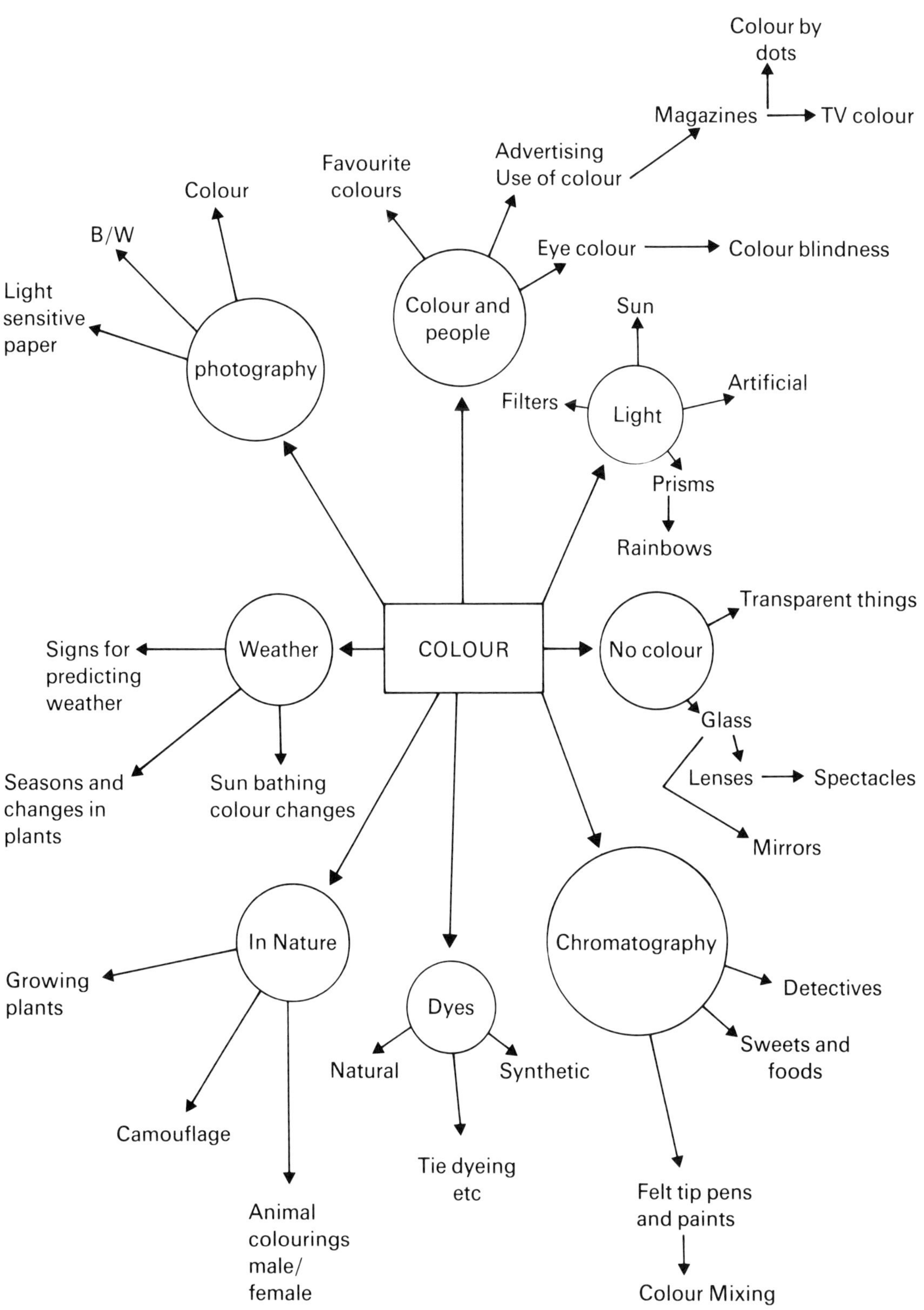

Colour by dots
Magazines
TV colour
Favourite colours
Advertising
Use of colour
Eye colour
Colour blindness
Colour
B/W
Light sensitive paper
photography
Colour and people
Sun
Filters
Light
Artificial
Prisms
Rainbows
Signs for predicting weather
Weather
COLOUR
No colour
Transparent things
Glass
Lenses
Spectacles
Mirrors
Seasons and changes in plants
Sun bathing colour changes
Growing plants
In Nature
Camouflage
Animal colourings male/female
Dyes
Natural
Synthetic
Tie dyeing etc
Chromatography
Detectives
Sweets and foods
Felt tip pens and paints
Colour Mixing

Some Ideas for Further Work on Colour

1 Colours in nature and camouflage.
2 Grow cress on moist cotton wool in identical containers covered with different transparent colour filters.
3 Use a magnifying glass to look at the way coloured dots are used to make up the coloured pictures in magazines.
   You make a picture of coloured dots and see what they look like from different distances.
4 Which colours are the easiest to see from a distance? Cars that are black or white, and black on yellow for number plates.
5 Weather sayings . . . sky at night, shepherd's delight.
6 Lights through combs and cracks.
7 Light-sensitive paper experiments.

PART 2

MAKING, EXPERIMENTING
AND PROBLEM SOLVING

CONTENTS, PART 2

About Part 2

This part builds upon the technical principles and guided experiments of Part 1. It is inclined to be a little more open ended and requires more thinking and manipulative skills. All the activities have been extensively tested in schools with a wide range of pupils.

The materials for each section are listed at the start of the section but the materials are generally to be found around the home, in the classroom or inexpensively purchased at the local Do it Yourself shop.

Preparation for the Activities

Before embarking on these activities with a class of pupils it might be worthwhile to gather together all the materials you require. This might be in the form of separate kits or as a collection of materials common to all the activities. I personally find the latter more useful as it allows the pupils a degree of choice of materials needed. To avoid a pile of materials to be searched through, I use a 'roly poly' storage kit. In each compartment I put different tools and materials and this can easily be seen and stored at the end of the lesson.

List of possible materials
Collection of cotton reels
Collection of different lengths of dowel rods (usually $\frac{1}{4}''$, small enough to go through the centre of the cotton reel)
Grommets (rubber washers either bought from Radio Spares, or made by slicing pieces of suitable plastic or rubber tubing. Should fit tightly on the dowel rods), or wind elastic bands around the rods a few times.
Off-cuts of wood and peg board
Rubber bands (assorted)
Pieces of card (stiff) various colours and sizes
Pieces of wood or fibre board as bases for models (stored separately from kits)
Lengths of wire (coat hangers)
Sellotape
Double-sided sellotape
Blu-tack
Paper clips (of various types including paper rivet type)
Springs (various types)
Plasticine
String
Drawing pins
Length of 30 amp copper earthwire from power main cable (DIY shop)
Plastic bottles (from which are cut plastic strips and washers)
Small electric motors (4·5 volts from toy shops)
Batteries (4·5 volts)
Bulbs of the correct voltage (4·5), bulb holders
Wires for connecting batteries and bulbs (use only insulated wire)

Tools
Scissors
Small hacksaws
Sharp knife
Small hand drill
Wire stripper
Measuring rule
Protractor

Concepts covered in this section
Gears
Friction and sliding
Circular motion
Off-centre pivots
Centripetal, centrifugal force
Gravity and falling objects
Release mechanisms
Motion, linear, forwards and backwards
Distance/time relationship
Electric motors and simple circuits
Elastic power and energy release
Twisting mechanisms
Use of different materials
Measurements and graphs
Measurement of angles
Simple electrical circuit, on/off switch
Conduction of electricity through metals
Insulation of electricity by plastic tape
Vibration of mass on wire
Rotation and interconnection of rotating drums
Reflection of angled mirrors
Dissolving
Soaking and absorbency
Mixtures of dyes and colour mixing
Separation of dyes by chromatography

LET'S MAKE A MODEL GO-KART

You will use some of the principles you have already looked at when studying wheels and mirrors in earlier experiments.

Constructional Details

Check that you have all of these
4 cotton reels
1 battery (4·5 volts)
1 motor (4·5 volts) with a pulley attached
1 piece of card
2 lengths of plastic coated electrical wire
2 wooden rods (dowel to go through the centre of the cotton reel)
2 grommets (or elastic bands). (This is a piece of rubber used to hold the wheels on)
1 rubber band

You will also need
Blu-tack
Sellotape
Scissors
Ruler
Pencil
Screwdriver

Here is a drawing of the model you are about to build:

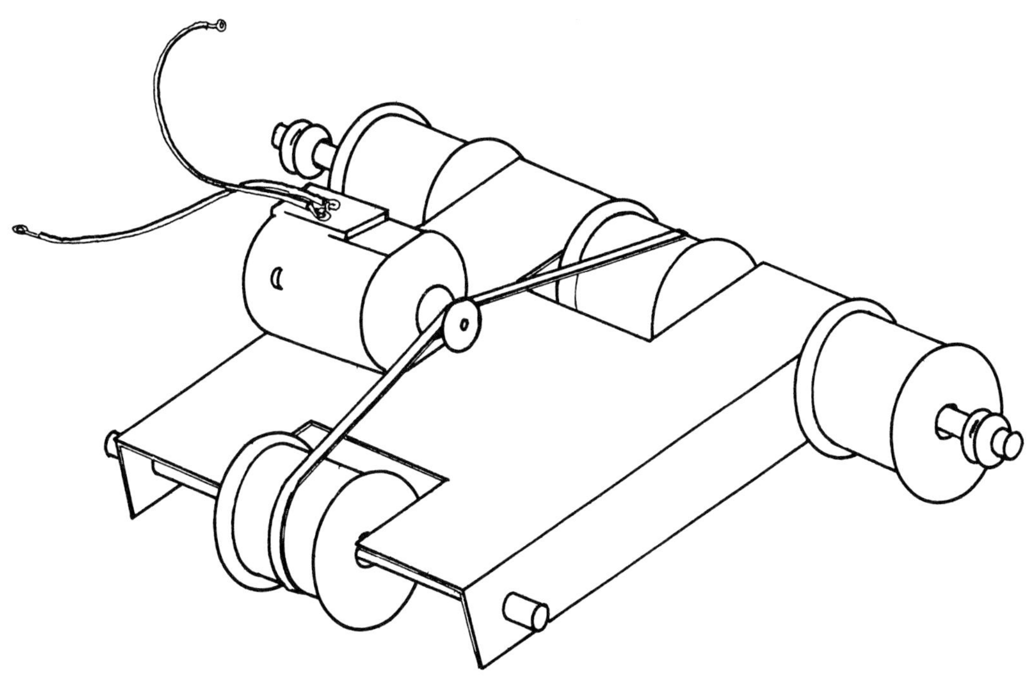

Take the piece of stiff card and follow the instructions carefully.

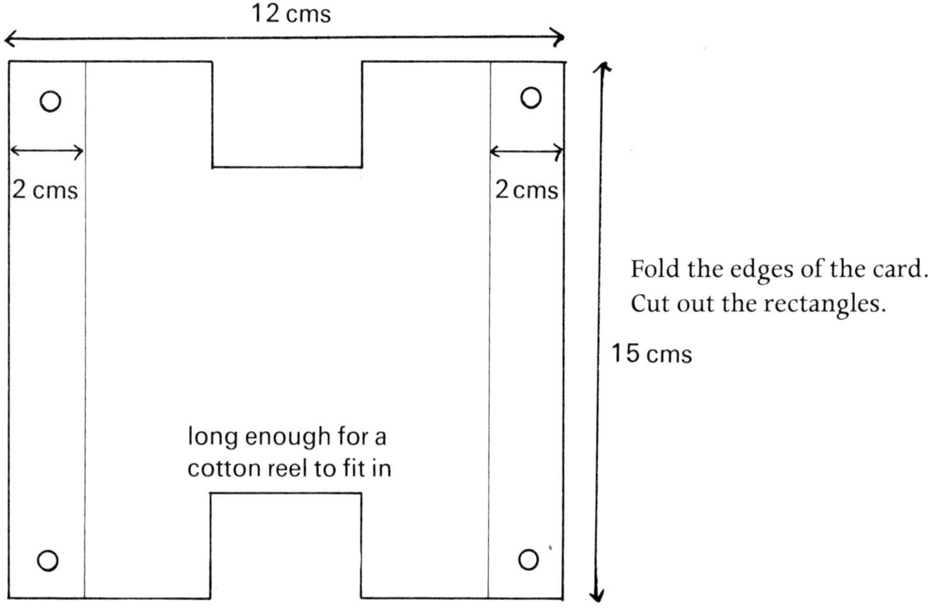

Fold the edges of the card.
Cut out the rectangles.

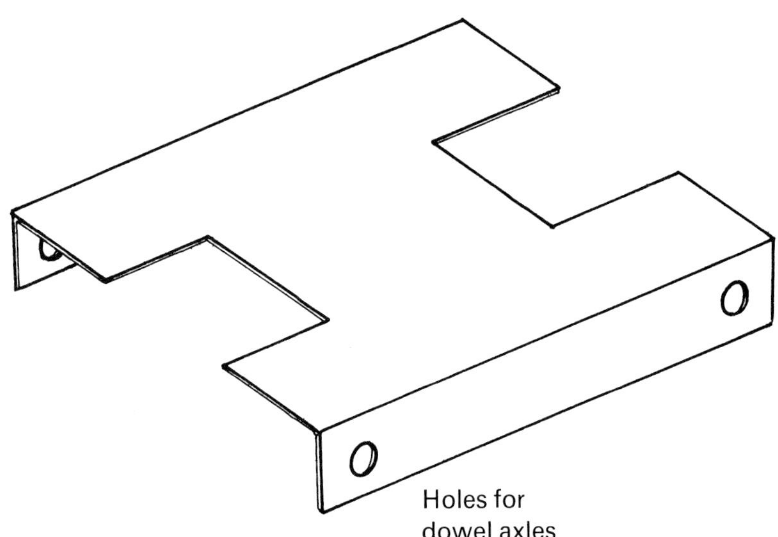

Hold a cotton reel in one of the cut out rectangles.
Push the short rod through the holes in one end of the card and through the centre of the cotton reel.

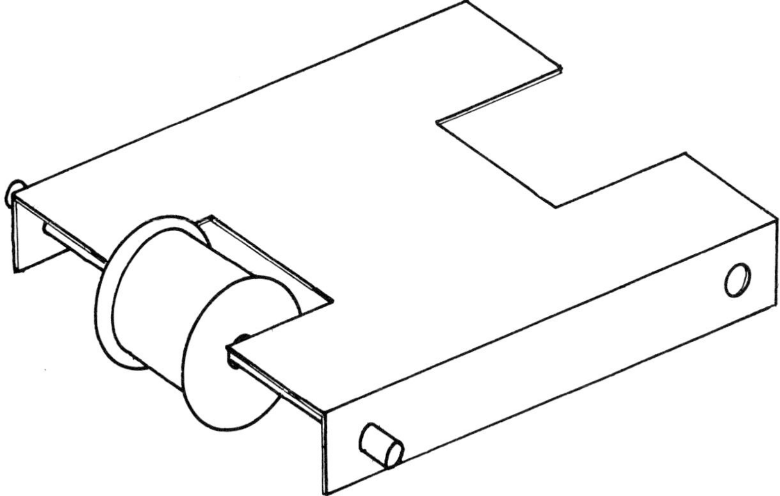

Give the cotton reel a spin to make sure that it turns freely.

Do the same at the other end of the card but using a long rod.

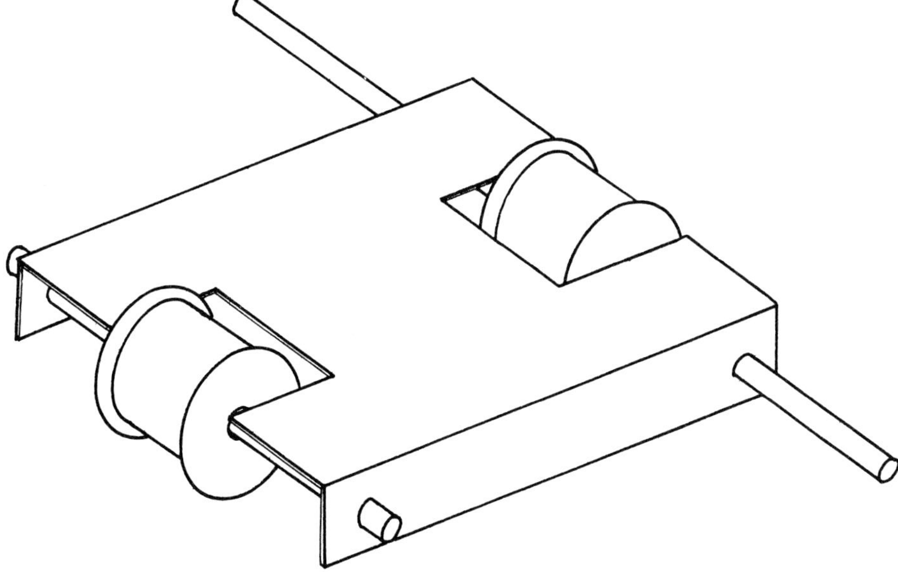

Place a reel on each end of the long rod.
Hold these in place with the rubber grommets (or elastic bands wound around a few times).
Make sure that these cotton reels turn freely.

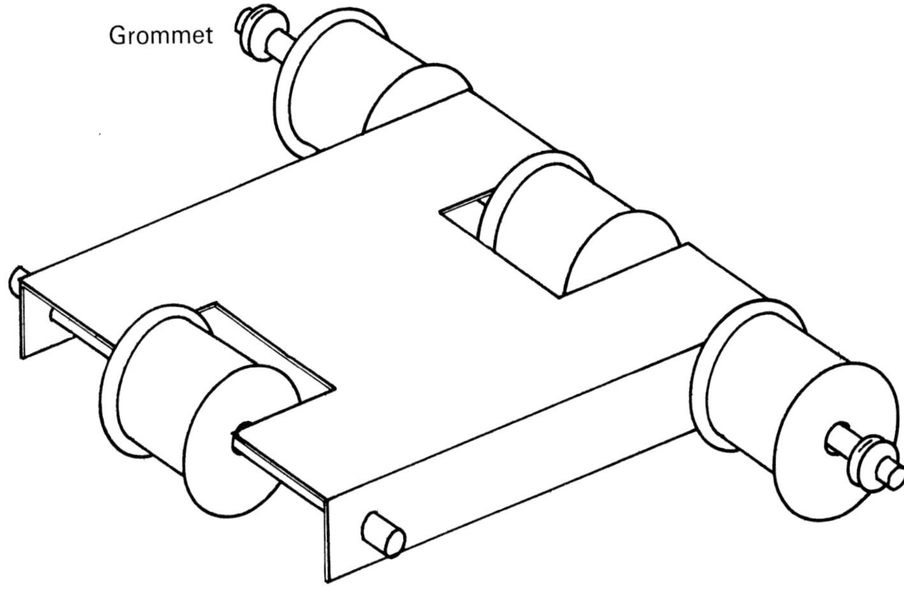

Stretch a rubber band round the two middle cotton reels

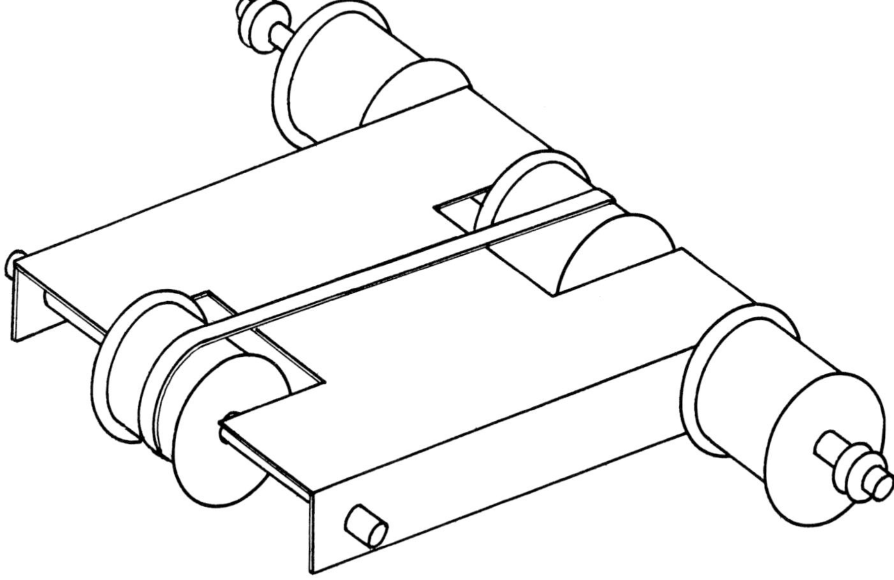

Stick the motor firmly in place with the Blu-tack.
Put the rubber band over the pulley.

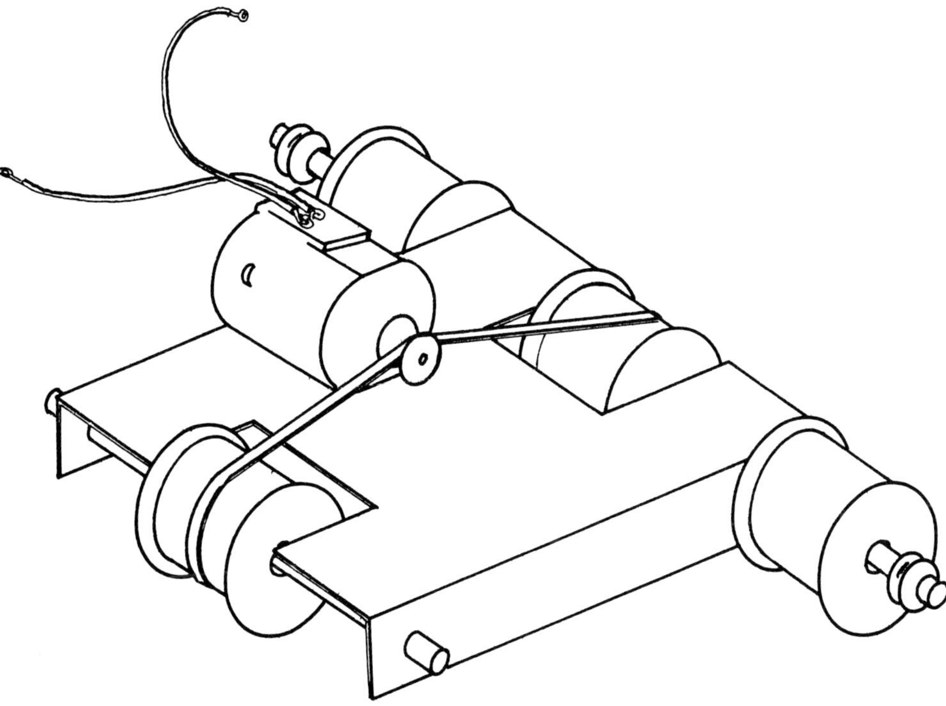

Stick the wires from the motor on each end of the battery with sellotape to make it go.
You can hold the battery or stick it to your go-kart with Blu-tack.

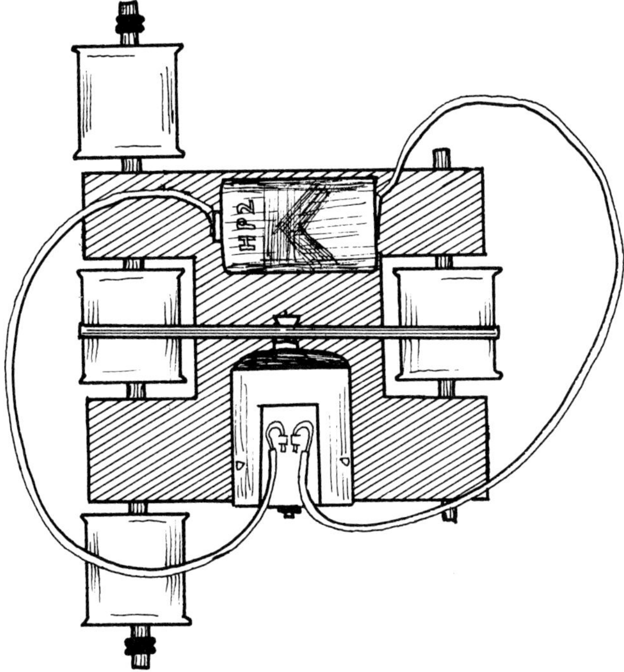

Your Four-Wheeled Car
It is unusual for cars to have their four wheels arranged with one at the front and three at the back. List some of the advantages and disadvantages of this arrangement for the four wheels on a car.

Problems For You to Solve

1 How can you make the vehicle go in the opposite direction?

2 Using the same battery, see if you can fit a light bulb in its holder to your vehicle so that it lights when the vehicle is moving.

3 How long does it take for the vehicle to travel a distance of two metres?

4 From your answer to question 3, can you work out how long it would take to travel:
   (a) Twenty metres?
   (b) One hundred metres?

5 What would you need if you wanted to make the go-kart turn to the left or right?
   How could this be done? Could it be done with two motors?

6 Can you work out the speed of your go-kart in metres per second?
$$\text{Speed} = \frac{\text{Distance (in metres)}}{\text{Time (in seconds)}}$$
Your answer will be in metres per second.

7 Draw a picture of your vehicle.

8 Make a car 'body' from paper or card for your vehicle.
   You could paint it to make it look like a racing car.

9 If you have a chance, study how an electric wheelchair works and how it can be controlled using electric motors.

How safe is the driver of your go-kart?
Make a seat for the driver of your go-kart from stiff card and make the model of a driver out of plasticine to fit the seat.

Stick the seat firmly to the car and place your driver in the seat.

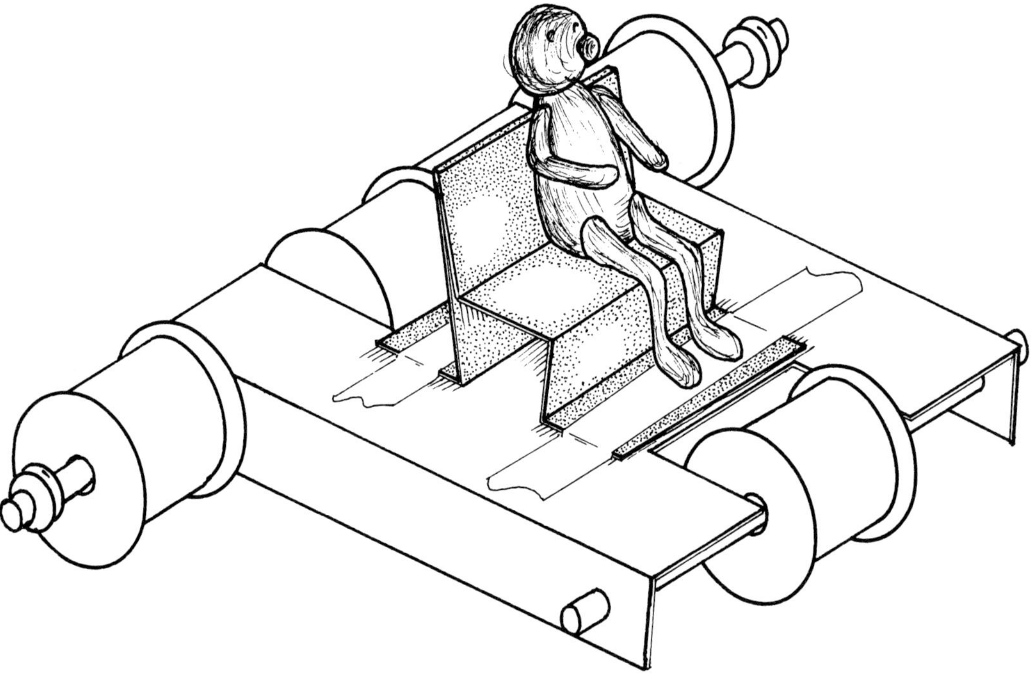

Run the car into a wall and see what happens to the driver.

Now put an elastic band around the driver as a seat belt and repeat the experiment. What happens now?

How could you make your go-kart more safe for the driver? After all, he might have been you!

Extension

STREAMLINE THE KART

Use the kart made in the previous experiment and do some tests on streamlining.

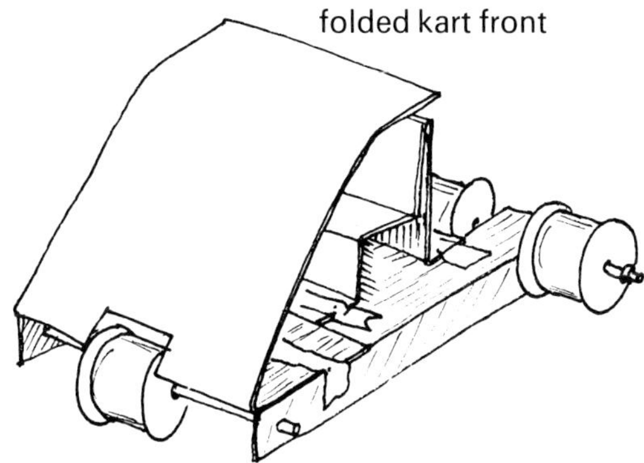

Make a series of folded cardboard fronts for the kart and run the kart down a wooden or cardboard slope (without motor attached).

From the same point each time predict which kart would go the furthest along the flat floor afterwards. Now try.

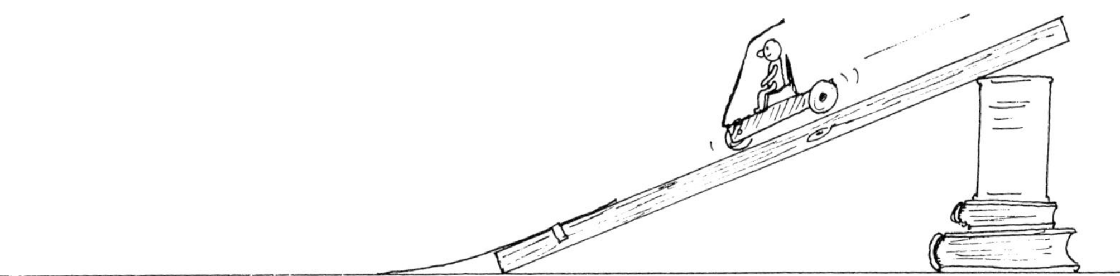

100

Could you say which shaped kart has the best streamlining?

You might have to repeat the experiment a few times for each design, and take an average of the results.

Why is this necessary when doing experiments involving measurements?

Which should be the most streamlined?

Does streamlining save petrol?

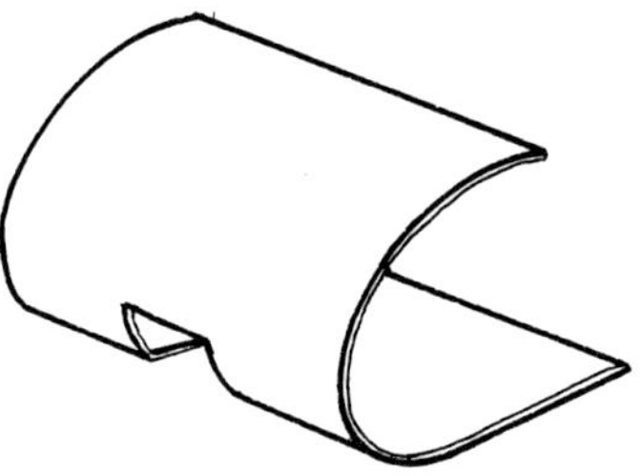

Possible folded car shapes for front of kart

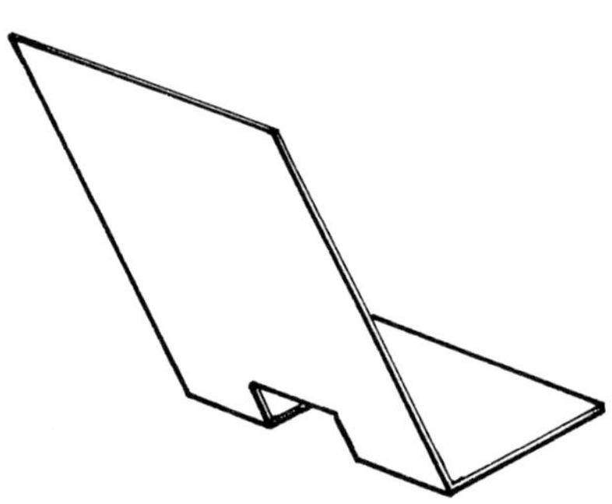

You could try the same experiment with toy cars using different streamlined fronts. Are lorries streamlined? Have a look the next time you are out to see if any large lorries with trailers have a flap fixed on top of the cab.
What do you think this is for?
Are buses streamlined?
Design and draw a bus shape which could be streamlined.
Are racing cars streamlined?
What advantages do streamlined racing cars have over other cars of the same engine size?
Why do racing cars have spoilers?

Mirrors on cars
If you wanted to fix a mirror on your go-kart to allow the driver to see behind him, where would you fix it?

THE TWISTER MACHINE

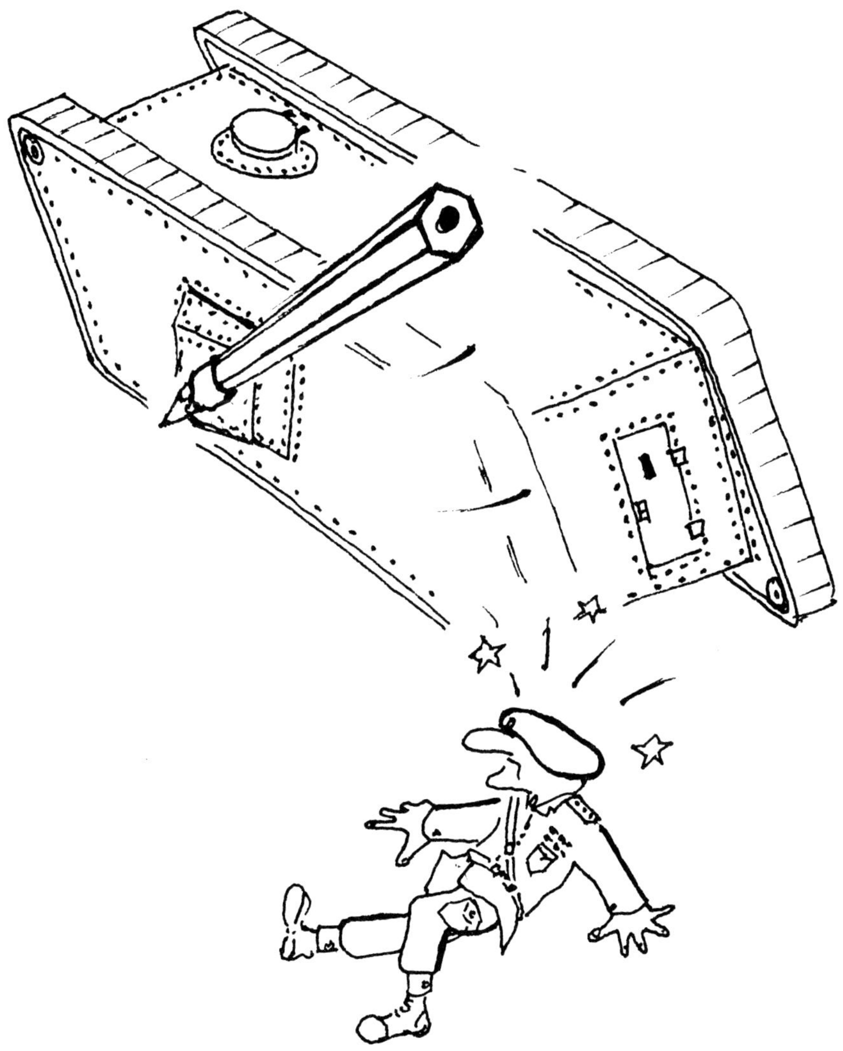

When making the twister you will be using some of the principles previously used in Part 1 when investigating elastic power.

Constructional Details

Check that you have all these things
Plastic bottle
3 elastic bands
Wooden rod (13 cms long) or long pencil
Pencil
Plastic washer (disc of plastic with a hole in the centre)
Plasticine

You will also need
Scissors
Rule

Cut the base off the bottom of the plastic bottle.

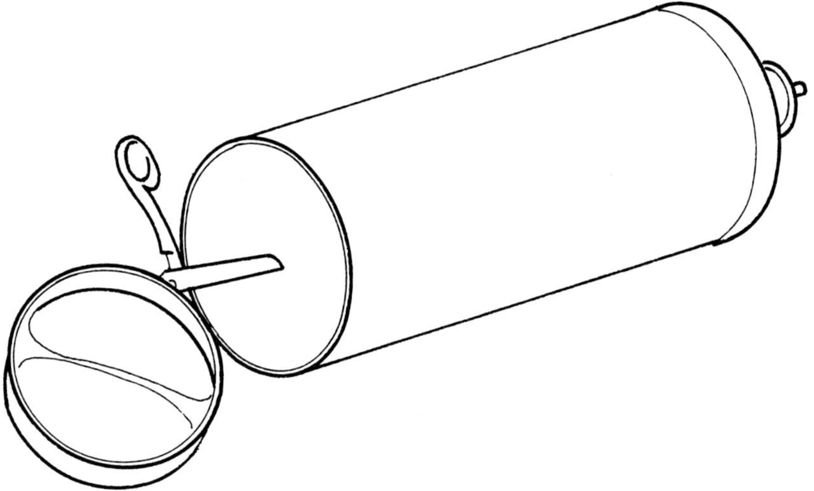

Put an elastic band around the bottle, 10 cms up from the edge you have cut.

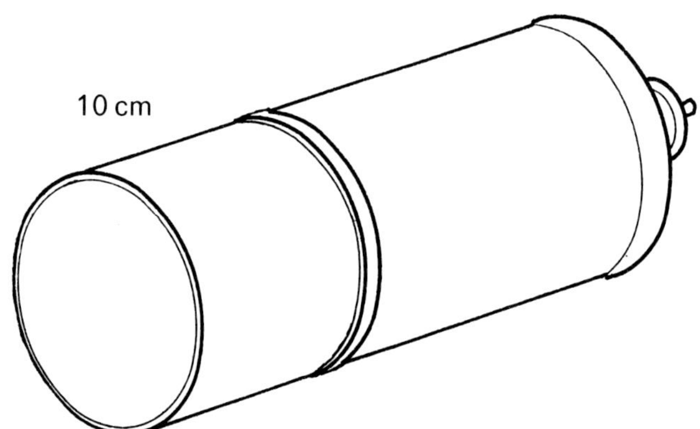

Using scissors, cut straight upwards as far as the elastic band.

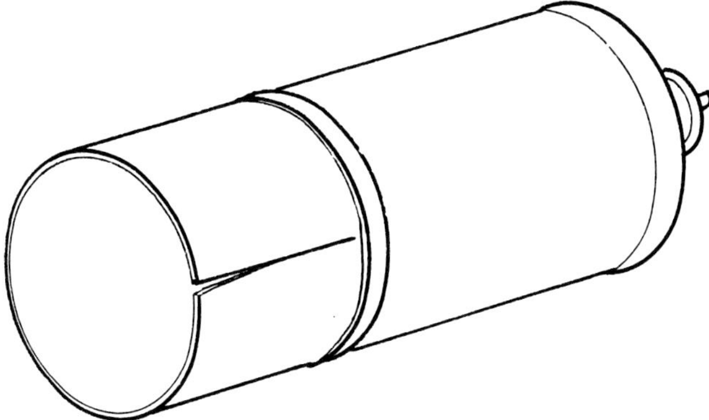

At 2 cm intervals along the bottom edge, make more cuts to form strips. This should be done all the way around the bottom edge.

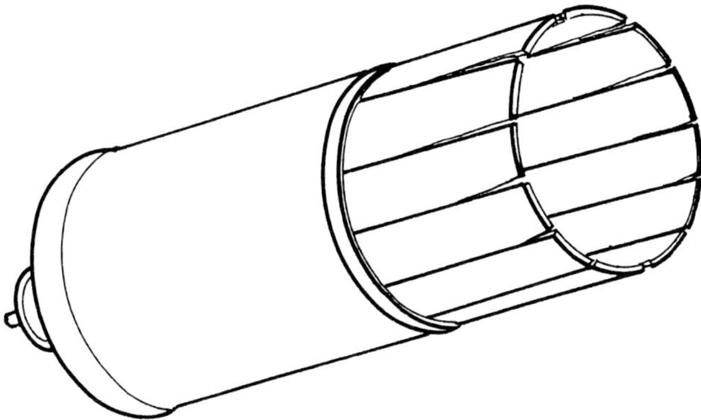

Cut the plastic cap off the top of the lid.
This will make a hole in the lid just big enough to squeeze a pencil through. If it is not, use scissors to make it a little larger.

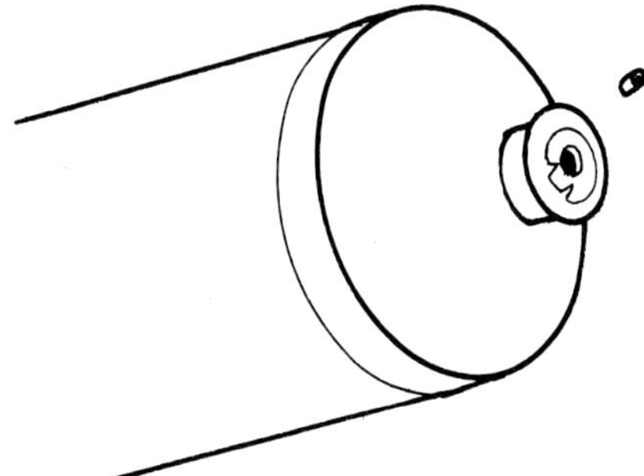

Make two holes in the side of the plastic bottle.
They must be 13 cms from the top of the lid, and opposite each other.

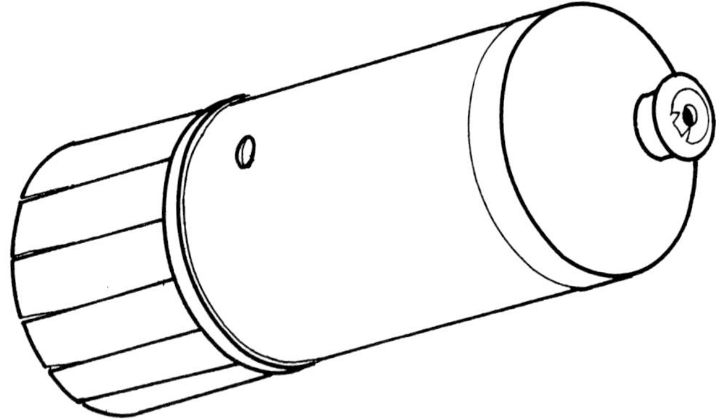

Push two elastic bands through the hole in a circular plastic washer, cut from a spare piece of plastic bottle.
Use a pencil to stop them from going all the way through.

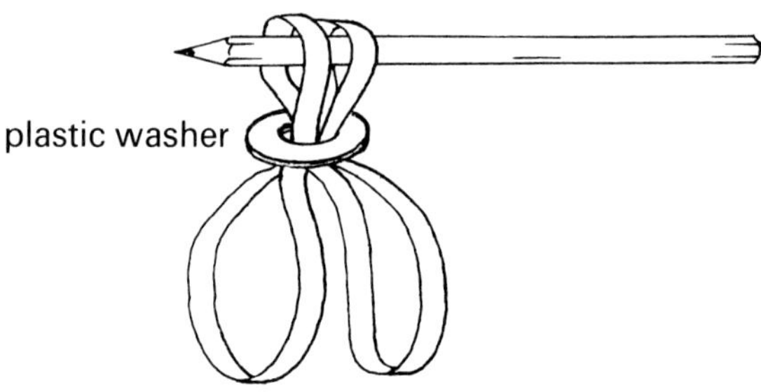

Thread the two elastic bands down through the hole in the lid.
The pencil should stop them from going all the way through the lid and plastic washer.

Push one end of the wooden rod through one of the holes in the side of the bottle, through the elastic bands and through the other hole.

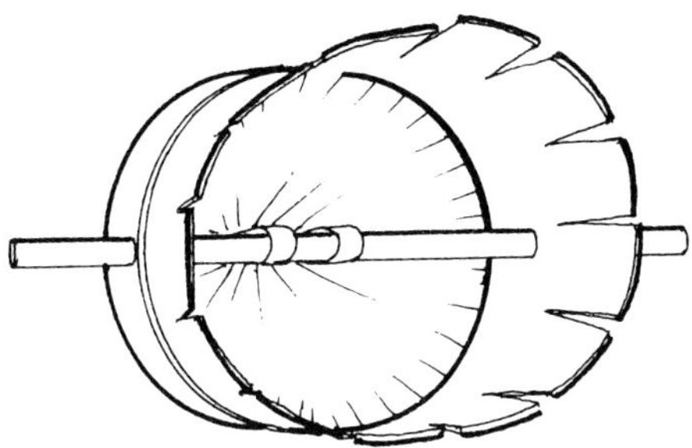

The rod now rests over the middle of the plastic bottle.

Curl the ends of the strips over and tuck them under the elastic band on the outside of the bottle.

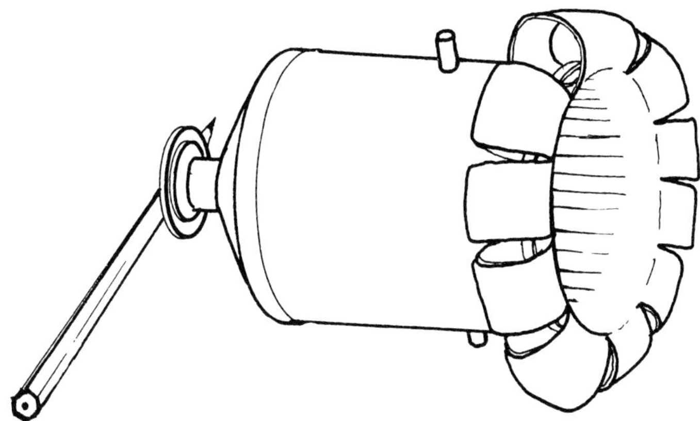

Turn the pencil round and round, so that the elastic bands wind up inside.

Problems for You to Solve

1   Use the pencil to make ten twists in the elastic band.
Place the Twister onto the floor as shown in the last diagram.
Let go of the pencil and see what happens. How many turns did it make?

Unwind the elastic band and start again.

Twist the band 20 times. Count the number of circles the Twister makes when it is put onto the floor.

Unwind the elastic band and start again.

Do it again for 10, 20, 30, 50, 60, and 70 twists but *not* for 40 twists.
Record each result.

10 twists goes ___________ turns

20 twists goes ___________ turns

30 twists goes ___________ turns

50 twists goes ___________ turns

60 twists goes ___________ turns

70 twists goes ___________ turns

*Remember to unwind the elastic band after each go.

2   Plot the results from question 1, on to your graph.

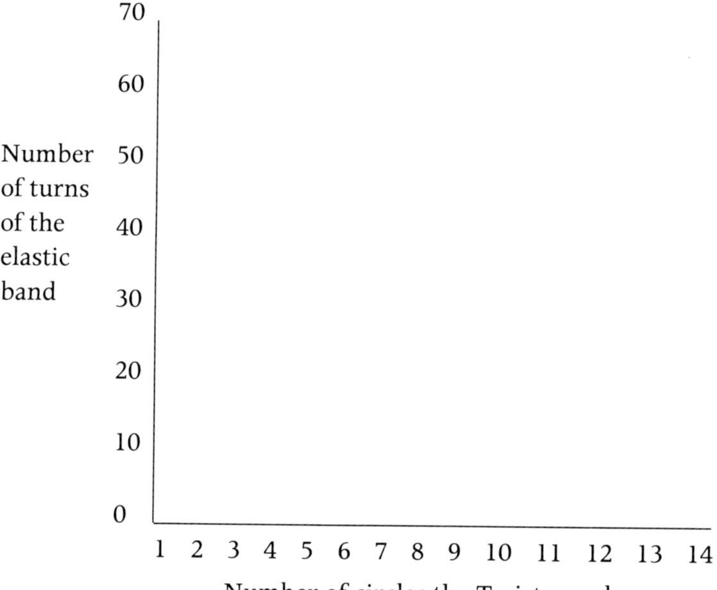

3   Look at the graph you have drawn.
    Estimate or guess the number of turns the Twister would make if the elastic
    band was twisted 40 times.

    Now see how good your guess was.

4   Your Twister machine travels in a circle.
    How could you make it travel in a straight line?

    Draw a picture of your design. Would it be worth making one to see?

5   Uncurl the ends of the strips and make the Twister stand upright.
    Wind up the elastic and watch how it moves.
    What effect do you think a small ball of plasticine would have if it was hung
    on a string tied on to the end of the pencil?

    Now try it and see.

ELASTIC POWERED JUMPING FROG

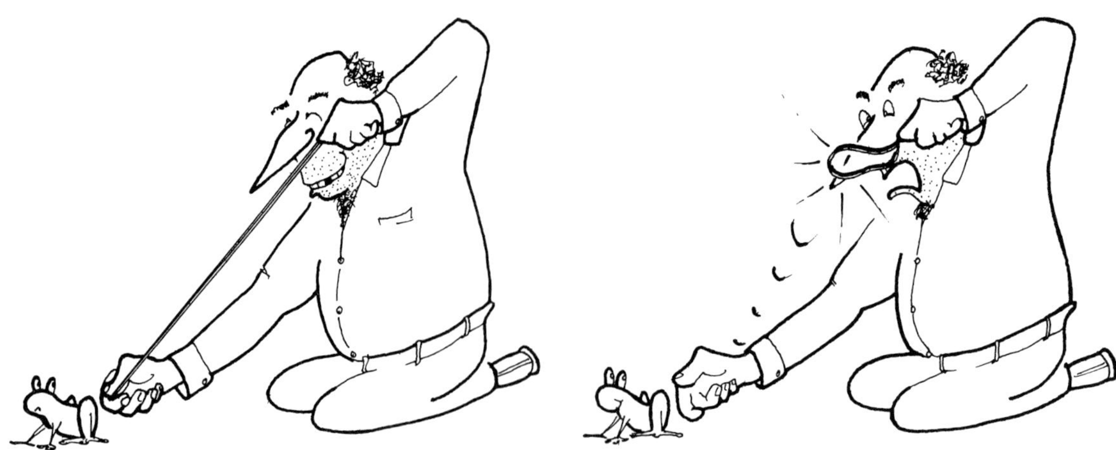

This will give you full instructions for making a jumping frog.

Constructional Details

Here's what you need
Piece of card
Paper clip
Rubber band
Protractor
Answer sheet
Hole punch
Masking tape

You will also need
Scissors
Metre ruler
Pencil

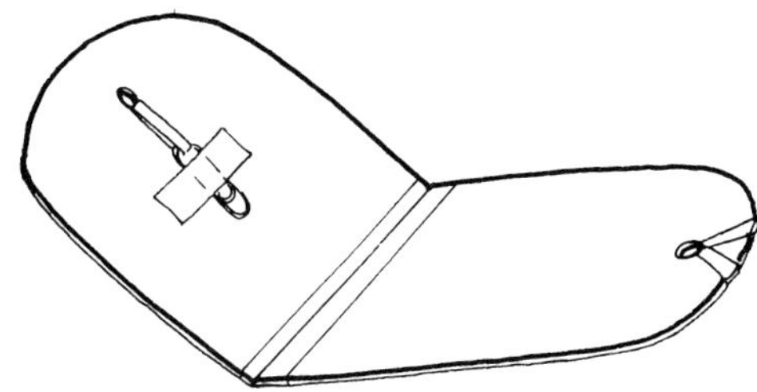

Here is a drawing of the model
you are about to build.

Take the card and make a hole at each end.
Now round off the corners of the card and draw a line across the centre and bend it backwards and forwards a few times.

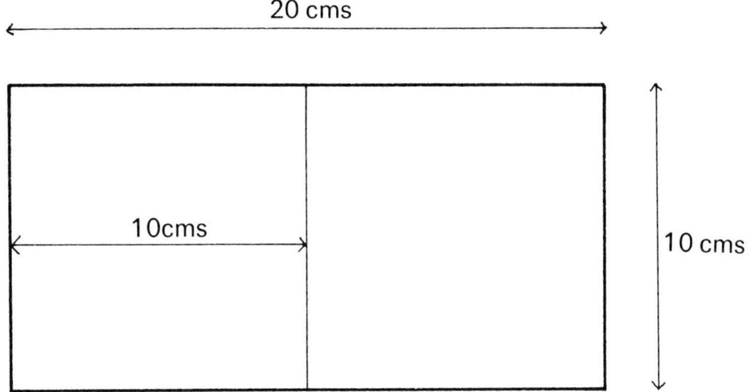

Push the rubber band through one of the holes and loop one end of the band through the other pulling it tight.

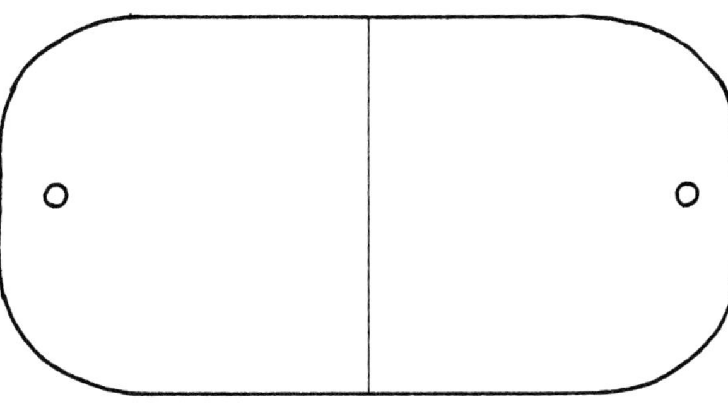

Push the loose end through the hole at the other end of the card.

Attach the end to a paper clip. Stretch the rubber band and fasten the paper clip in position with masking tape (see below).

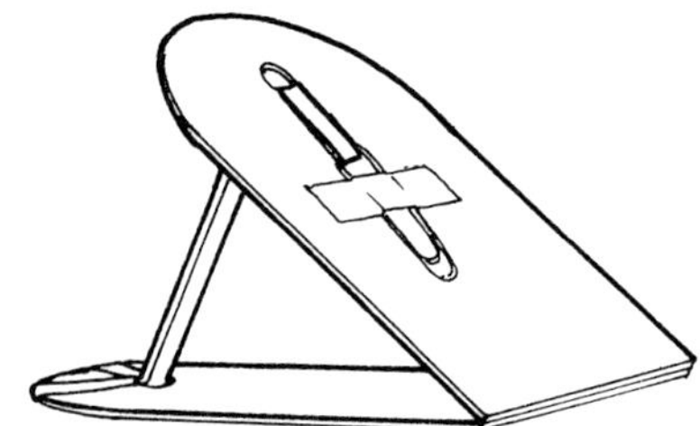

Bend back the flap of your frog so that the paper clip is on the inside and the band is stretched over the outside, and lay it flat on the table with your finger at one end.

Suddenly release your finger. Measure how HIGH the frog jumps. (Use a metre rule placed up against the wall).

Write your answer here: _______________

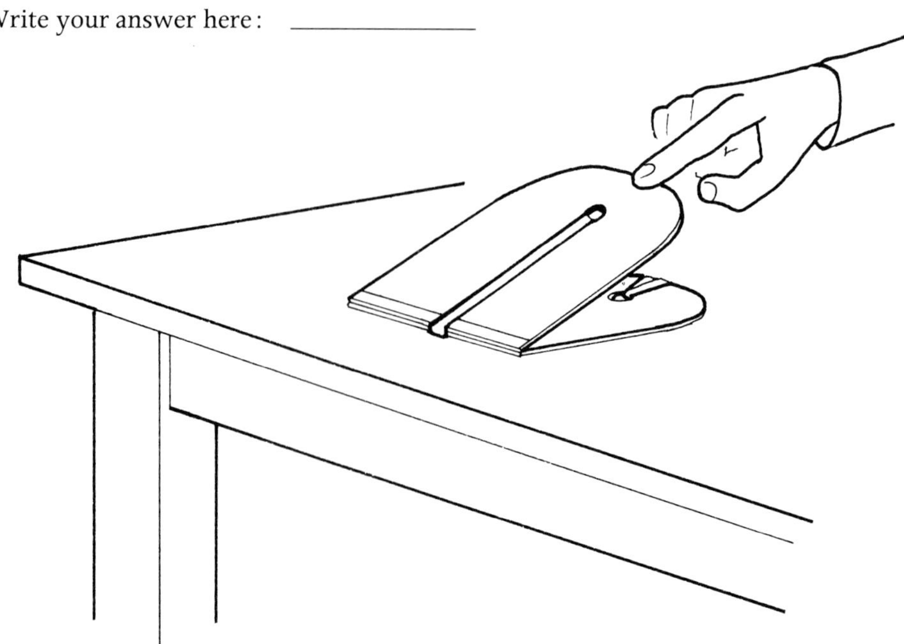

What makes your frog jump?

Now think of ways to make your frog jump higher. Write your ideas here.

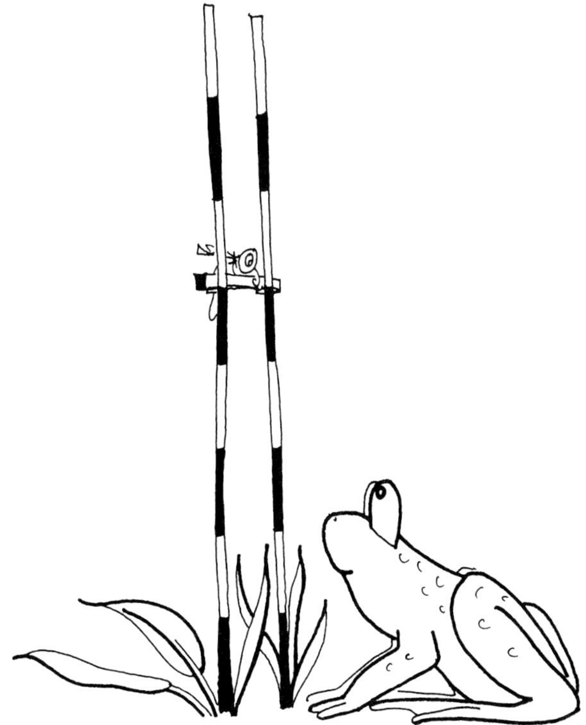

One way of making your frog jump higher is to shorten the rubber band. To do this, release the paper clip from the masking tape and the rubber band. Tie a knot in the rubber band. Fix the paper clip back in position, in exactly the same place as before.

How high will your frog jump now? Write your answers here.

Another way to make your frog jump higher is to make the holes nearer to the edge of the card.

By doing this (or by shortening the rubber band), you are making the band tighter so that it causes the frog to jump higher.

A Problem

Two children placed their frog on the floor in four different ways, separating the flaps at four different angles, and tried to find out whether it changed how high the frog jumped.

0° degrees (frog flat)	45°	90°	135°
Height in cms			

Put the results of these jumps here

Can *you* make a jumping kangaroo?

What causes real frogs and animals to jump is the stretching of muscles and releasing of this stretching energy, very much like an elastic band frog.

GRAVITY POWER

This section on the use of gravity builds upon the experiments earlier in the
book and poses some problems and possible solutions related to gravity.

Constructional Details

Here is what you need
Wooden base (20 cms × 20 cms) or peg board
2 cotton reels
Plastic washer (cut from the side of a plastic bottle)
Grommets (or elastic bands)
2 plastic strips (cut from the side of a plastic bottle) or 2 strips of stiff cardboard,
   10 cms long × 3 cms wide
4 wooden rods (three approx 12 cms long and one 4 cms long)
4 drawing pins
Plasticine
Cotton thread
Scissors
Sellotape
Rule
Pencil

Here is what to do
Take the wooden base and drill holes for the three 12 cm pieces of dowel as shown
below. (If you use peg board place the dowels in approximately these positions.)

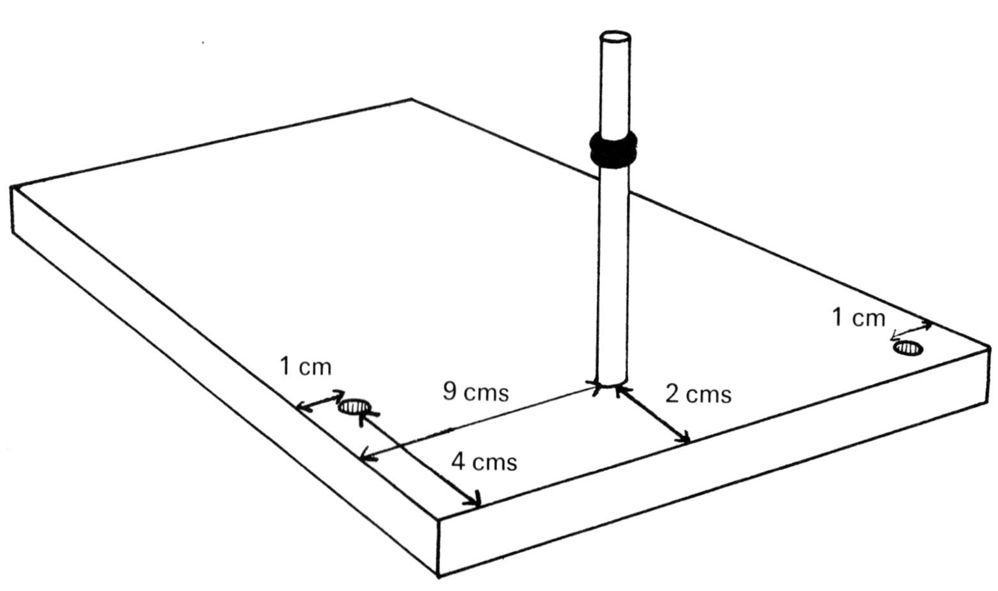

Put the grommet and plastic washer on the piece of dowel in the central hole as shown.

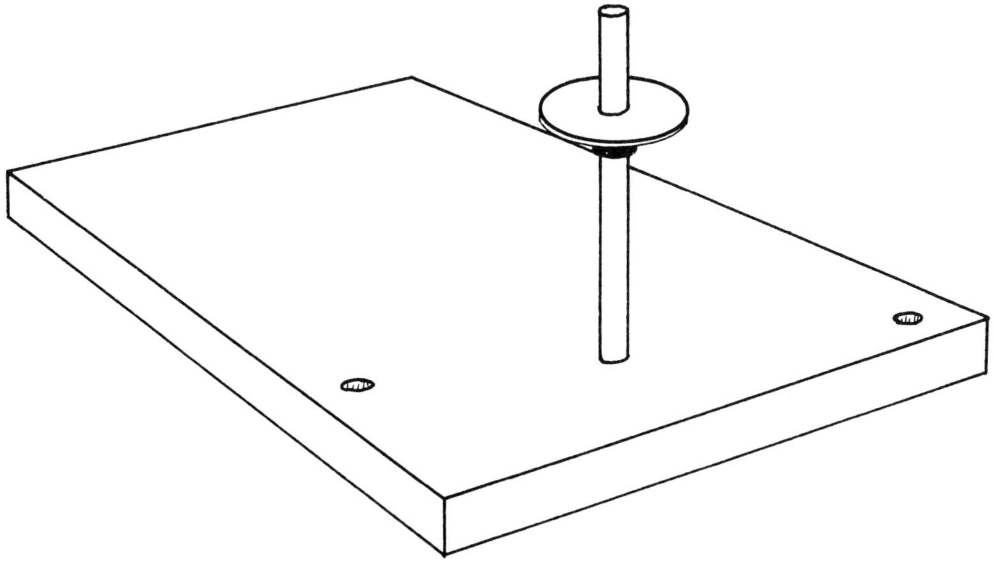

Make a hole in the side of one cotton reel and make a side arm on the cotton reel with the small dowel rod, like the one shown.

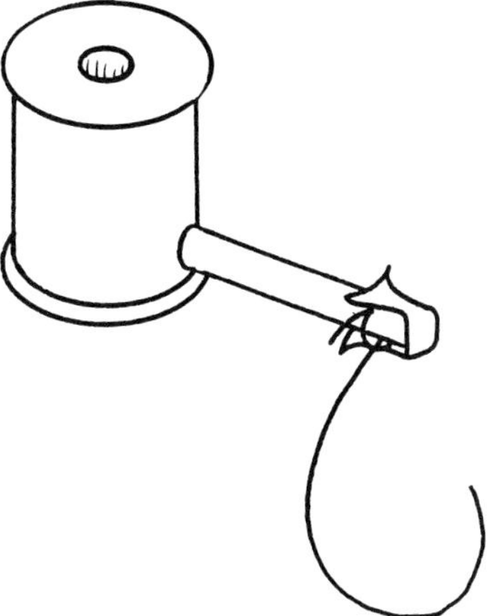

Measure and cut a piece of cotton about 12 centimetres long. Stick one end of this cotton to the end of the arm with sellotape. (This can easily be adjusted later if it is too long.)

Make a ball of plasticine (about the size of a small marble) and fix this to the other end of the cotton.

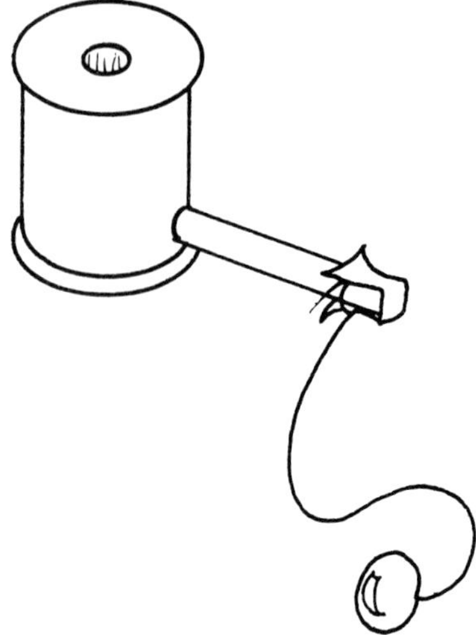

Slide the ball of plasticine along the piece of cotton to make a flying pendulum seven centimetres in length.

Take a longer piece of cotton (approx. 1 metre long) and tie one end to the rod arm *as close to the cotton reel as you can*.

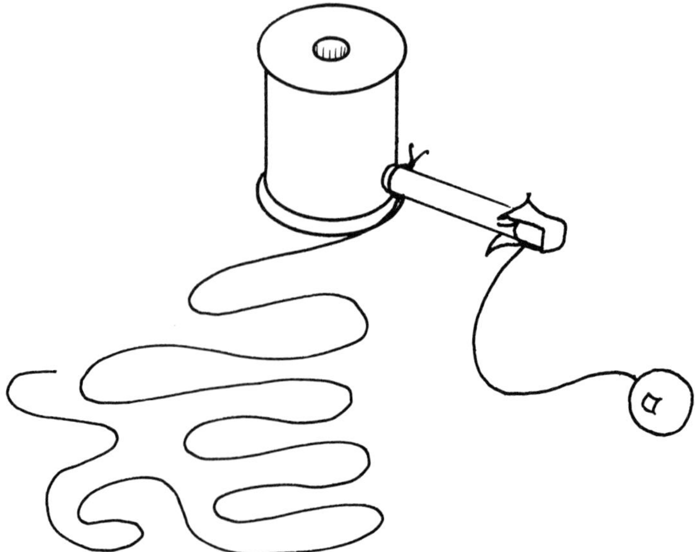

Now wind some of the cotton around the reel above the rod arm.
Make 18 turns then cut the cotton off.

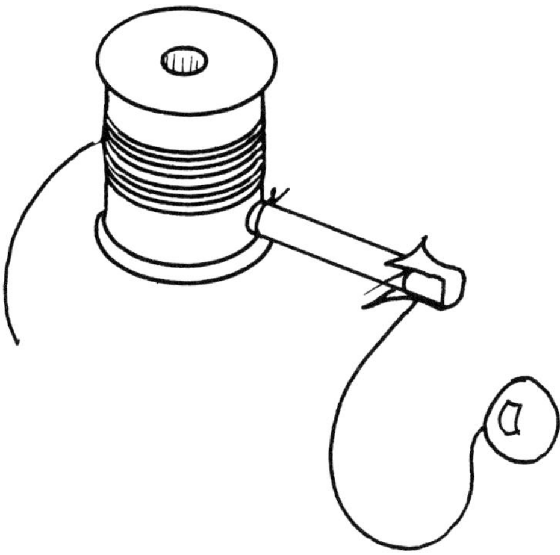

Make a ball of plasticine about the size of a large marble.
Fix this to the free end of the cotton.

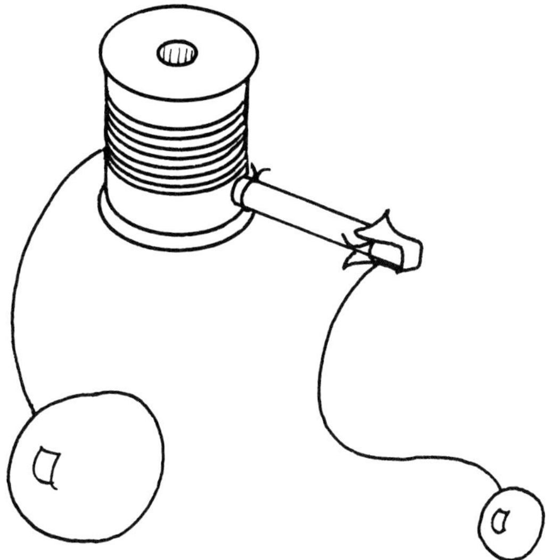

Check that the cotton reel is the correct way up (see diagram).
Place it onto the rod so that it rests on the washer.

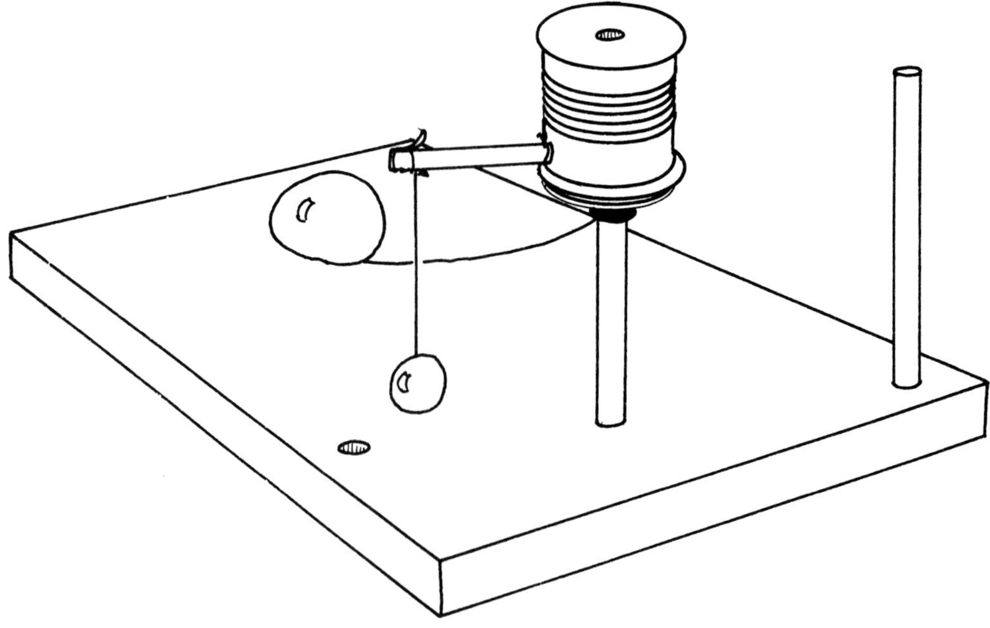

Take two pieces of plastic or stiff cardboard strip and cut a small hole in each,
1 centimetre from the end.
Draw a line across the plastic or cardboard strips at 3 centimetres from the end
without the hole.

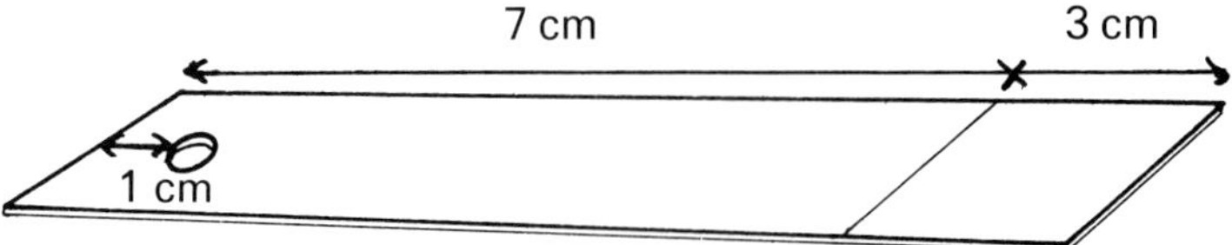

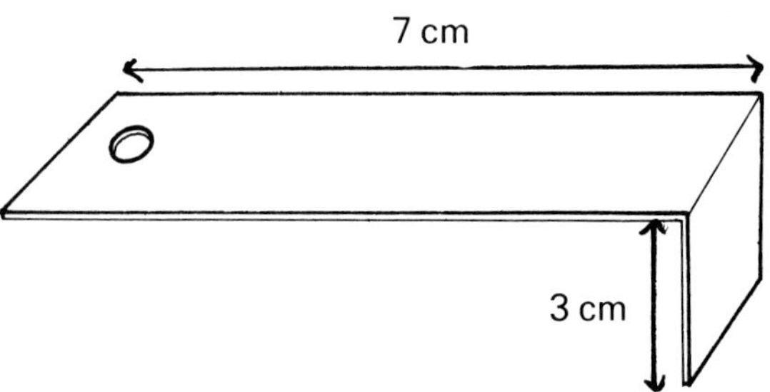

Place the strips on to the positions on the base, as shown below.
Fasten each of them in place with two drawing pins.

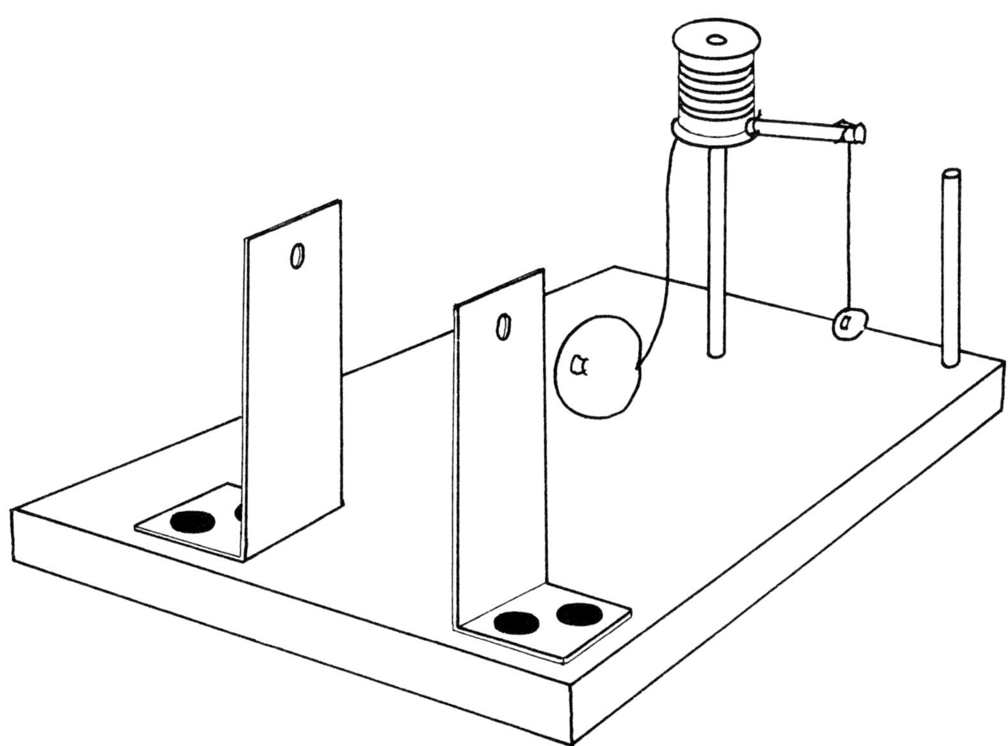

Push rod through the holes in the strips, holding the second cotton reel in the middle.

Place the larger piece of plasticine (and its cotton) over the roller-like cotton reel and gently let it fall to the ground.

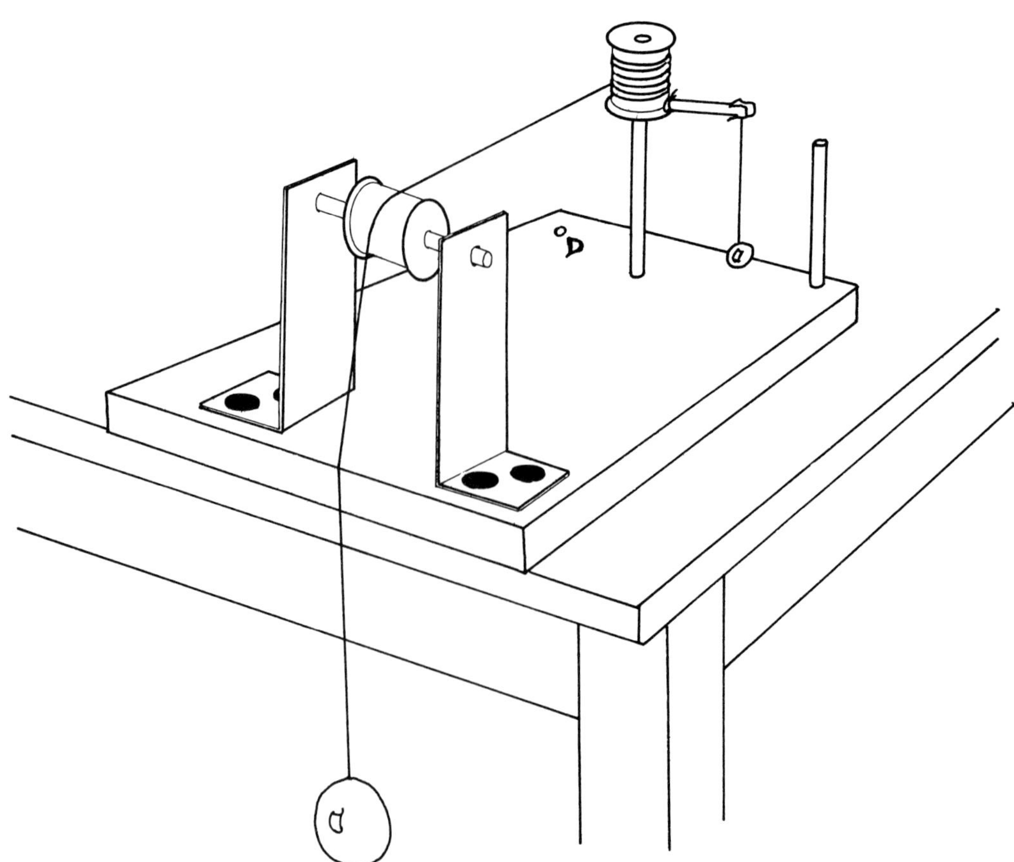

Notice

1   What happens to the cotton reel with the side arm?
    If the side arm rod and attached swinging weight doesn't twist around the upright rod, then adjust the length of the string.

2   Notice, as the heavy weight falls, the cotton reel and weight turns and stops, turns and stops.

Problems for You to Solve

1   How long does it take the plasticine ball to fall?
    How many stops does it make as it falls?

2   What do you think will happen if you place a second rod in hole D? (See last diagram.)
    Answer:
    Now try it.

3   If you alter the mass of your falling plasticine ball, will you alter the number of stops it makes?
    Answer:
    Now try it.

4   Why does this machine act like a clock?

5   Explain how the clock works by placing the following eight sentences in the correct order:

A   As the ball falls the vertical cotton reel turns.

B   This stops the falling ball.

C   The falling ball starts to drop again.

D   The flying pendulum wraps around the rod.

E   As the vertical cotton reel turns, the flying pendulum flies outwards.

F   The ball is released and starts falling because it is pulled by gravity.

H   The flying pendulum unwraps itself.

Other Ideas Using Gravity

1 Can you think of other types of clocks (other than pendulum clocks) that use gravity to measure time? (Hint—sometimes smaller ones are used as egg timers.)

2 How can falling water be used to turn wheels (water wheels)?
How can falling water be used to generate electricity?

3 A man wanted to reach the top of a mountain and he did not want to climb it? How could he use a machine which overcomes gravity to reach the top of the mountain? Are aeroplanes, balloons or helicopters, machines which overcome gravity?

4 Some power stations use cheap off peak electricity to pump water from a lower lake to a higher level lake during the night and then use the falling water power to help generate electricity during peak periods during the day. Dinorwic power station in Wales is one such modern power station which does this. It uses its gravitational force of falling water to turn generators to produce electricity.

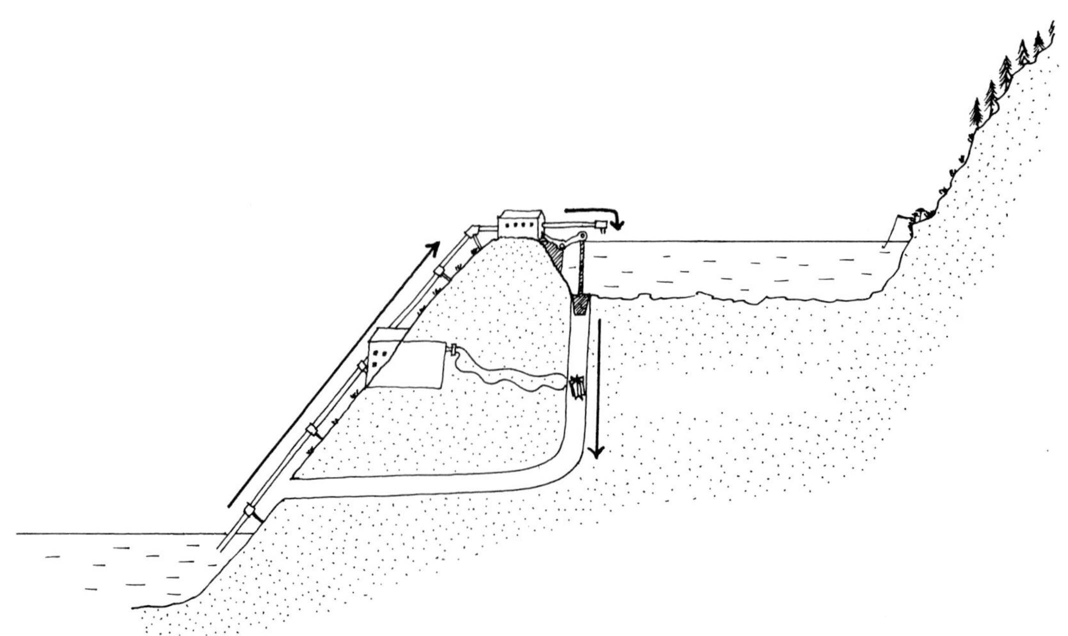

The Gravity Flasher

You will need
Wooden base (about 20 cms × 20 cms) or
peg board
2 metal rods (eg. from metal clothes hanger)
Clock with a second hand or digital watch
String
Plasticine
Bulb (1·5 volts)
Bulb holder
Battery (1·5 volts)
Paper clips
Copper wire
Rule
Pencil
Sellotape
Metal rod or clamp stand
Wooden or metal rod

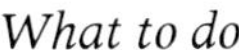

What to do

Wrap the copper wire around a pen or rod a few times. The spring you made must not be wrapped around too tightly. It must be able to slide along the rod. Make it this shape. Leave long lengths at each end.

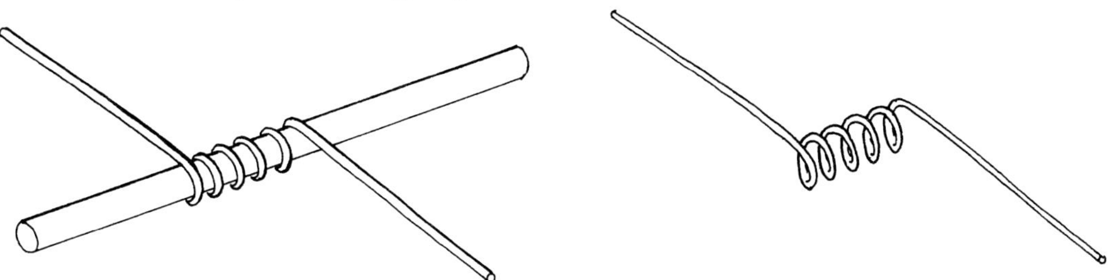

Attach a ball of plasticine to one end of the wire.

Now vibrate the ball and watch
what happens.

Reverse the spring on your rod
and try the experiment again.

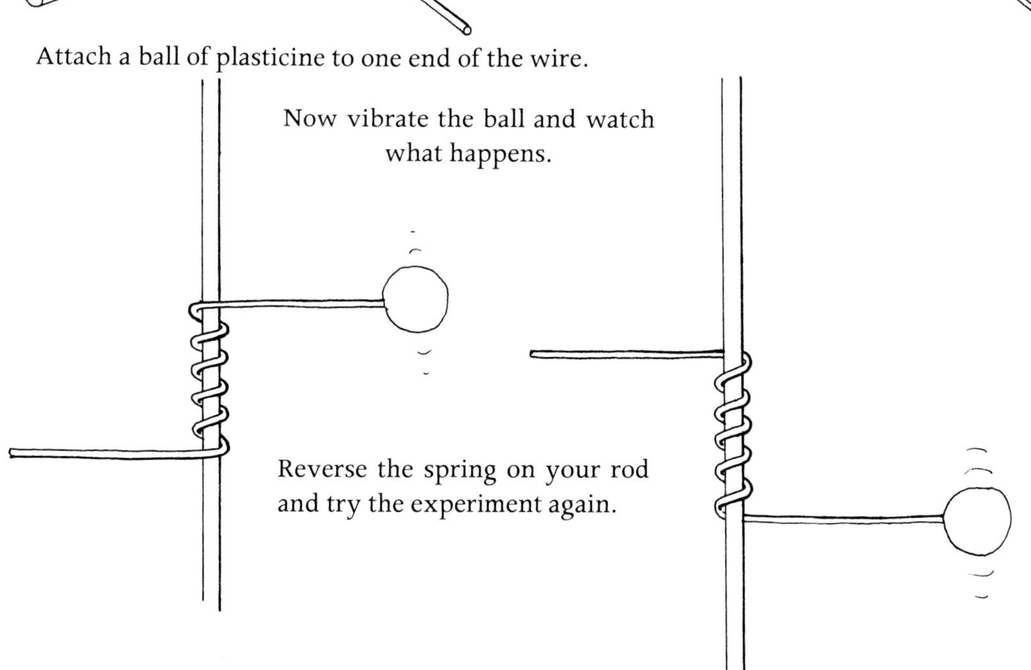

Repeat this experiment with the other springs with 3, 5 and 6 turns around the rod and complete the following table to show how long it takes the spring to reach the bottom of the rod.

SPRING (Number of turns)	TIME TAKEN IN SECONDS
3	
5	
6	

PROBLEMS FOR YOU TO SOLVE

Devise an experiment to find out if plasticine weights of different amounts affect the time take for the vibrating spring to fall.

Devise a system that uses the slow falling action of the copper wire spring to switch a 1·5 volt light bulb and battery on and off. This is a flasher circuit or gravity flasher.

There are a few hints below on making the gravity flasher, but try to work out the possible answer first.

Hints for gravity flasher circuit

HINT 1
Use a simple battery/bulb circuit.

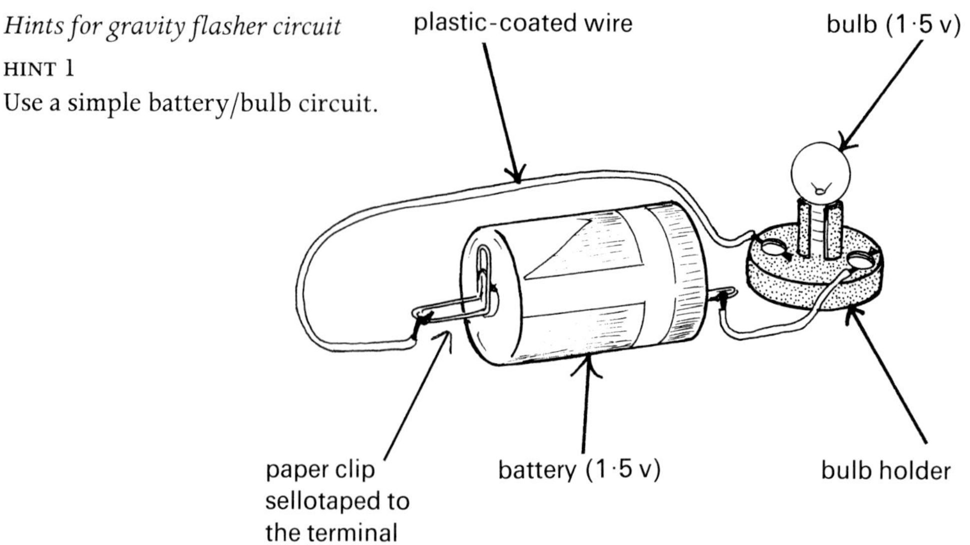

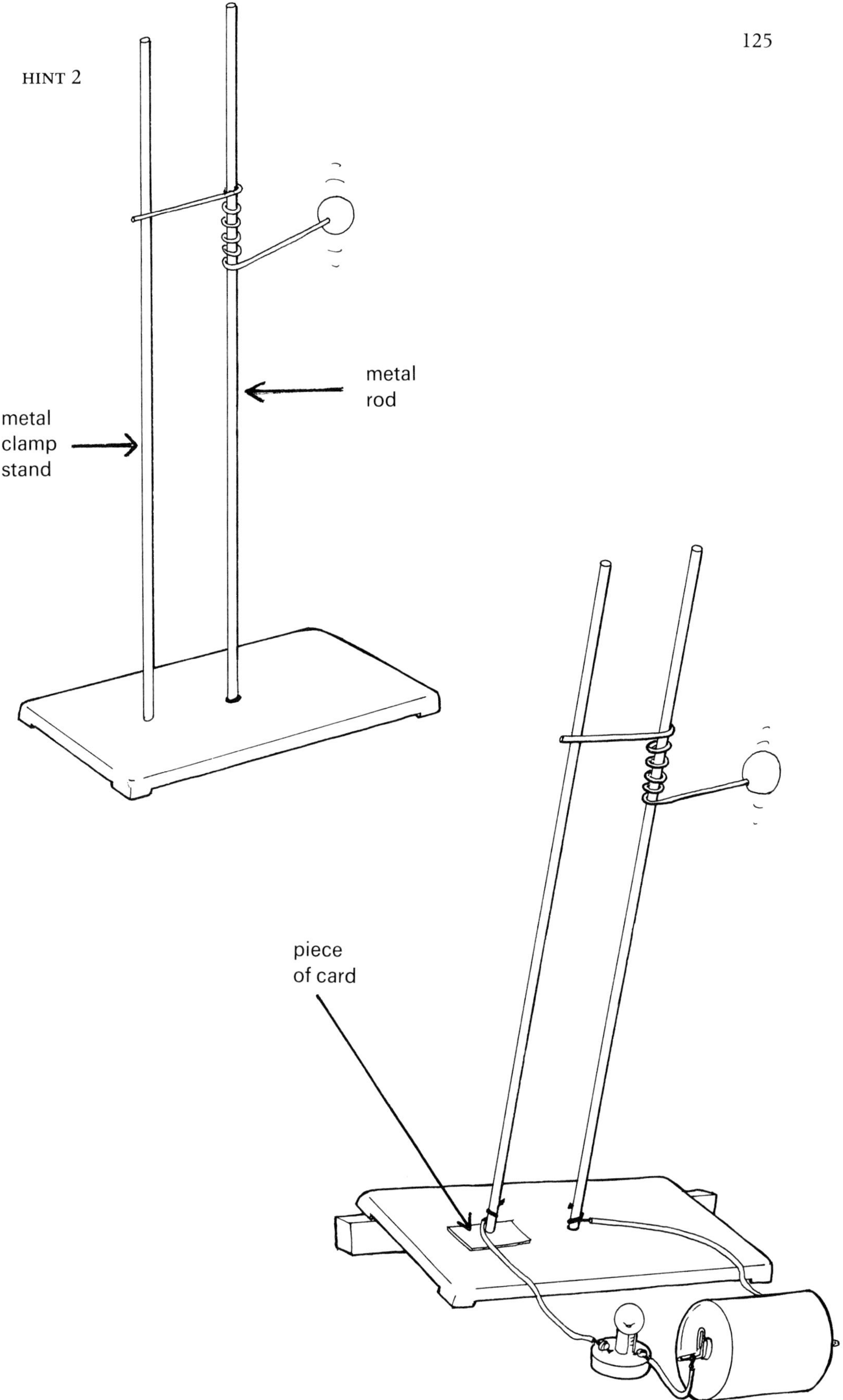
HINT 2
metal
rod
metal
clamp
stand
piece
of card

HINT 3

strips of
non-conducting
sellotape

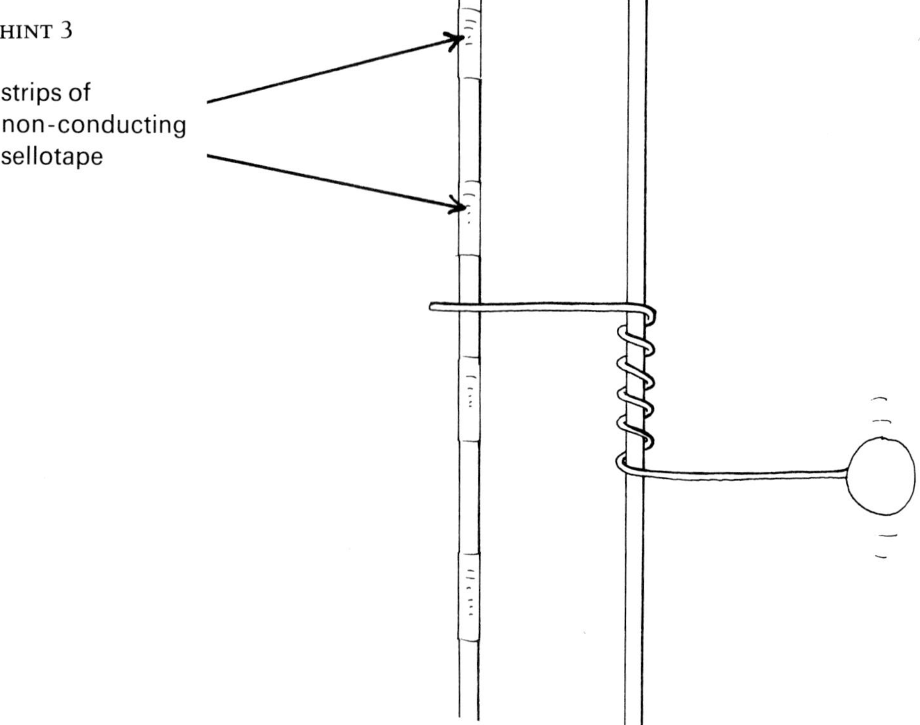

As the spring falls, does the light stay on all the time?
You might have to tilt one side of the stand to make a good contact between the wire and the clamp stand.

Can you see how the flasher works?

Other ways of using gravity to switch a light on and off

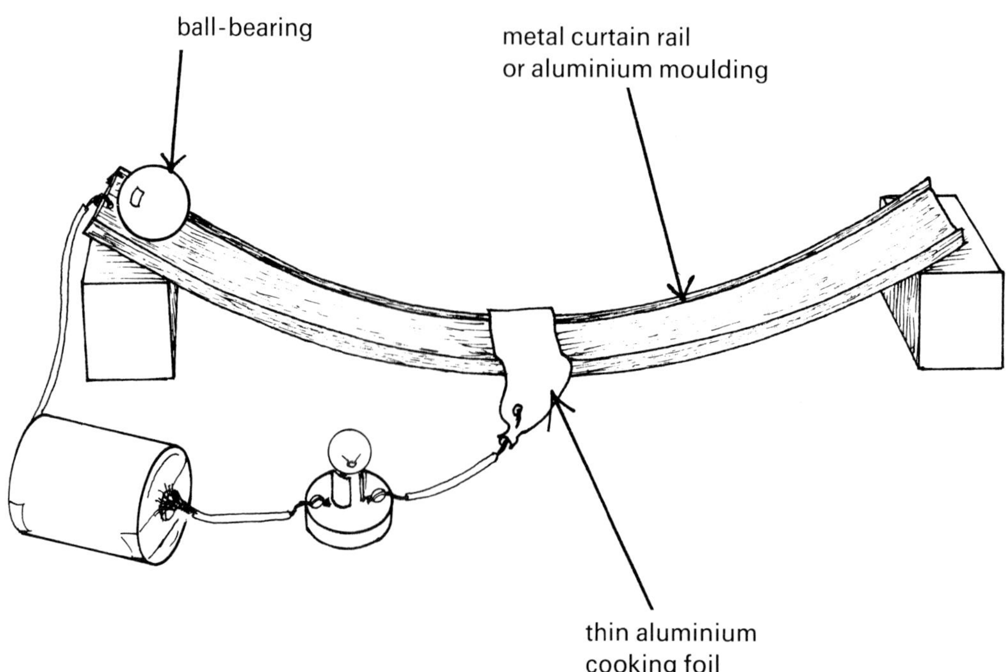

You can try a similar thing with cars on a track.

Can you think of other ways of using gravity to turn things on and off, eg. see-saws, etc. ?

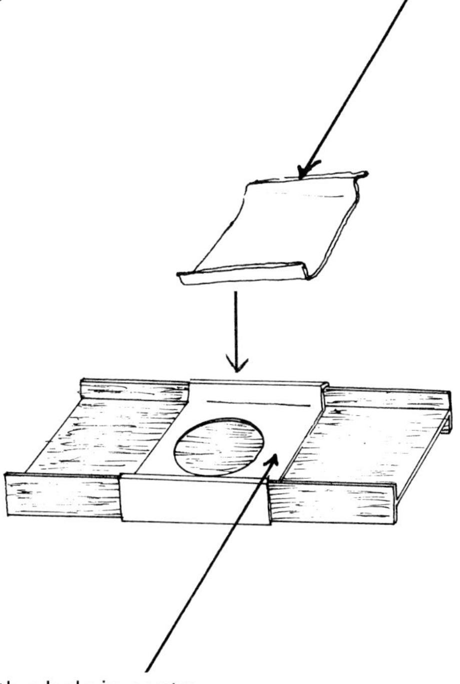

Piece of paper with a hole in centre.
This allows the ball-bearing to push the aluminium foil onto the metal base and so complete the circuit.

THE FLYING MACHINE

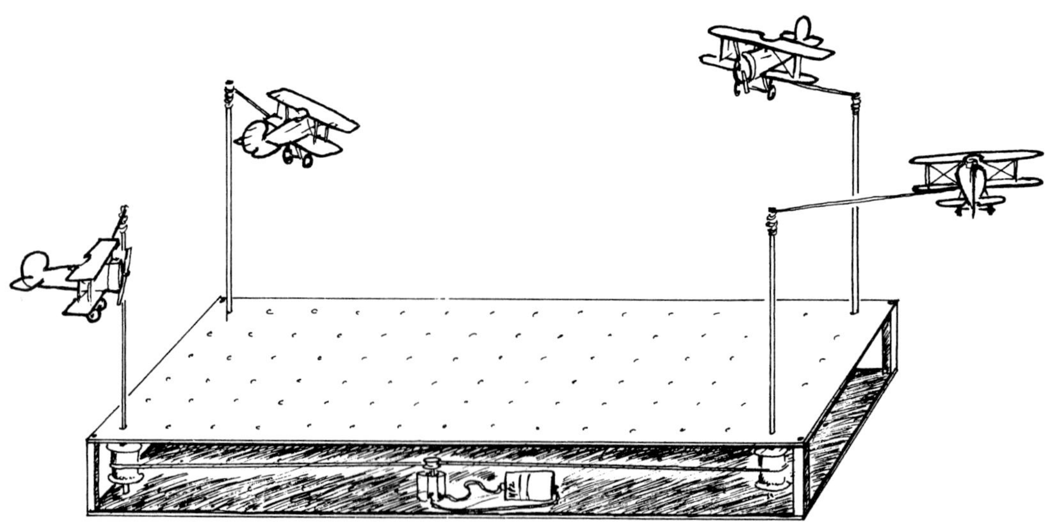

After a special outing of a class of pupils to the airport and a 30-minute flight over their town, the class wanted to make a moving model of flying planes. The whole class contributed ideas, artwork and construction skills to produce this excellent flying circus.

Constructional Details

You will need
1 square of plywood or blockboard as base (40 cms × 40 cms)
1 square of peg board as top (40 cms × 40 cms)
4 cotton reels
4 lengths of dowel (approx. 20, 25, 30, 35 cms long)
4 battens of wood for each corner (10 cms)
8 grommets and 8 plastic washers
1 electric motor—4·5 volts (preferably one which can be geared down, eg. Lego or Meccano motor)
1 4·5 volt battery plus plastic coated connecting wires (to match the electric motor)
4 lengths of stiff wire from coat hangers (10, 20, 30, 35 cms)
Elastic bands
Card to construct the planes

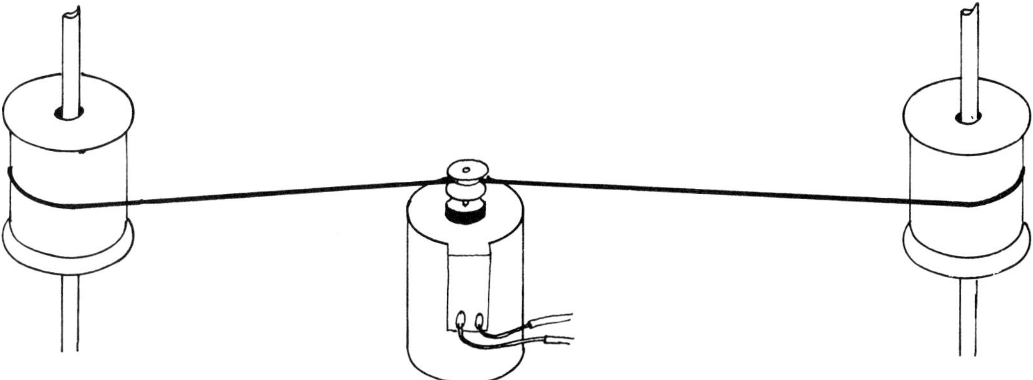

The dowel fits tightly through the cotton reels (this can be wedged with paper). The dowel rests in a hole in the base square of wood of the model.

Cotton reels rest between the two sheets of board separated by wooden battens at each corner.

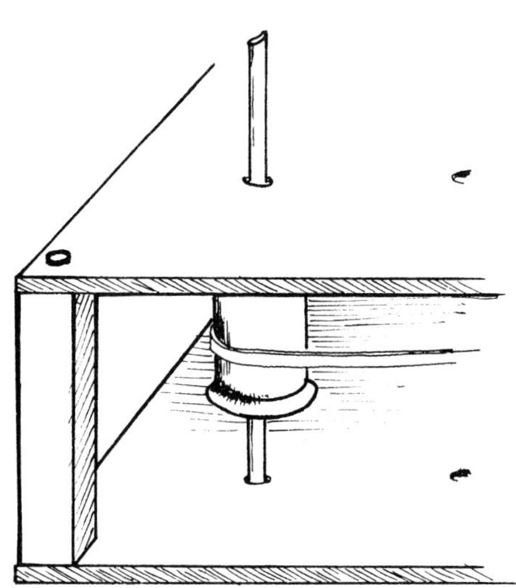

The elastic bands interconnect each adjacent pair of cotton reels and one reel is attached to the electric motor.

Sometimes the motor will not turn all the planes, but the Meccano motor with suitable gearing ratios works very well. You might have to modify the elastic driving mechanism depending upon what motor is available.

This is a great attraction when completed and is a great success on parents' evenings and when visitors come.

Instead of aeroplanes other roundabouts can be designed using the same principles.

MR FUNNY FACE

The principles of turning wheels and the use of elastic bands can be used to make some interesting and motivating machines for your hobby or classrooms.

This original model was used and devised by a class of pupils in a special school on the occasion of a visit of the health visitor who came to encourage the class to clean their teeth, and gave out free toothpaste.

Constructional Details

You will need
Cornflakes packet
Paints and plain paper
Empty tins as gears
Elastic bands
Dowel for axles
Plain card

Use all the principles you have learnt about elastic bands and wheels to construct the model as shown.

Mr Funny Face has eyes that move to and fro and an arm holding a toothbrush which cleans his teeth.

This mechanism moves by two tin can wheels turning back and forth. They are 'joined' by an elastic band.

The back and forth movements can be made by attaching the strips off centre to one side of the wheels.

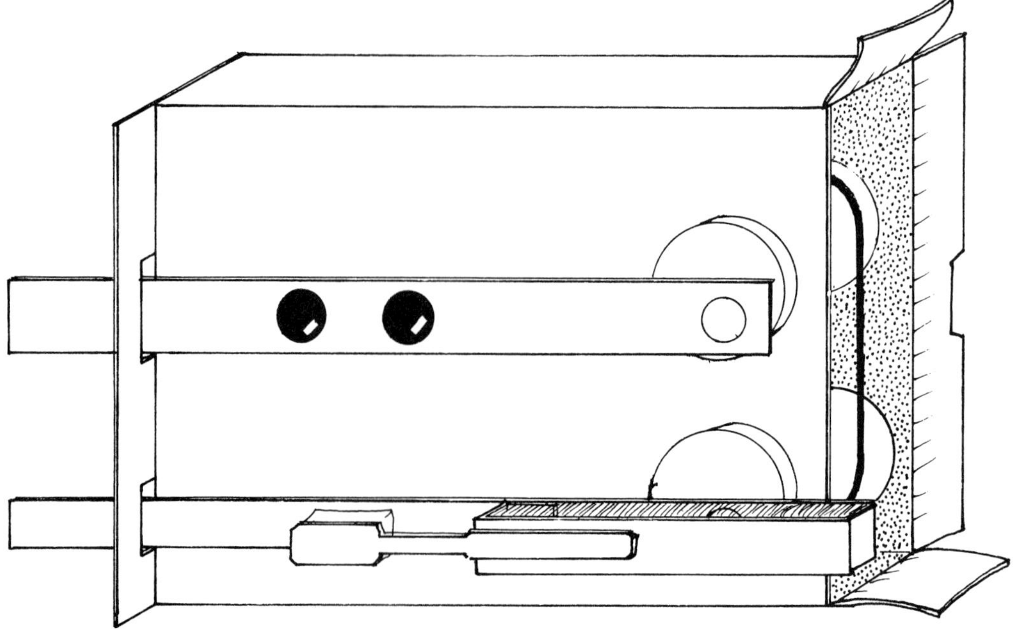

Paint a face on paper and stick to the outside of a cornflakes packet.

If you are clever you might also cause his moustache to wiggle as well!

MAKING A PERISCOPE

Looking over the top of people using a periscope!

134

You will need
Sheet of stiff card (from cornflakes packet opened up)
2 mirrors
Protractors or 45° setsquare
Scissors

What to do

1   Take your piece of card and draw three lines to divide it up into four equal
    widths.

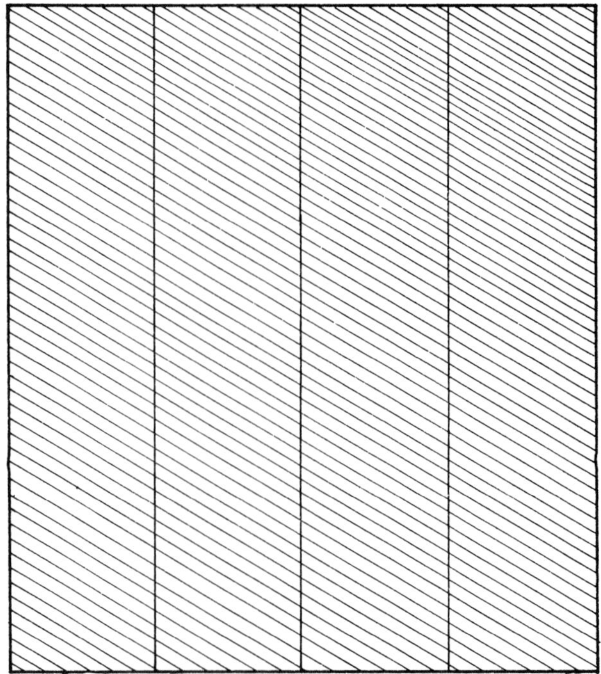

2   Fold the card along these lengths, just to show that they can be folded to
    meet each other.

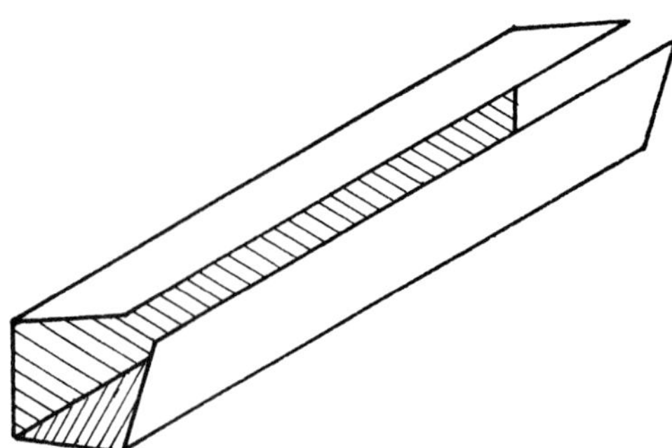

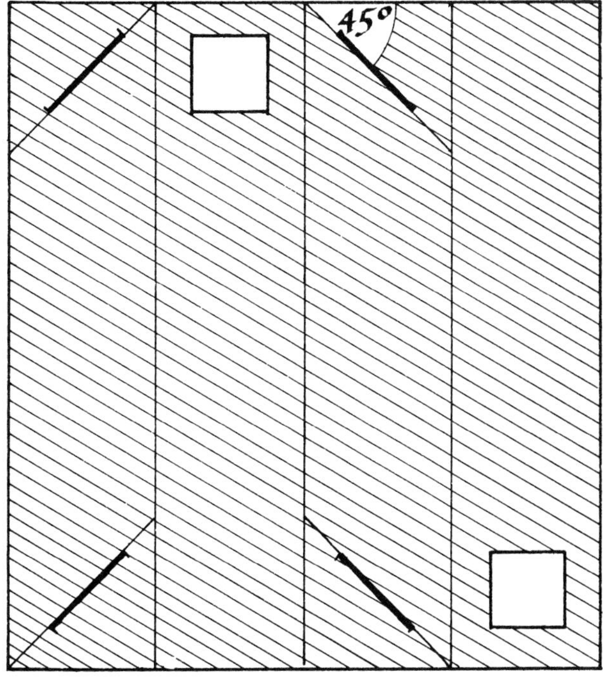

3 Open up the card again.

4 Cut out the squares in the positions shown.

5 Mark pencil lines at 45°.

6 Cut slits for your mirrors along the lines but try not to cut fully to the top and bottom edges.

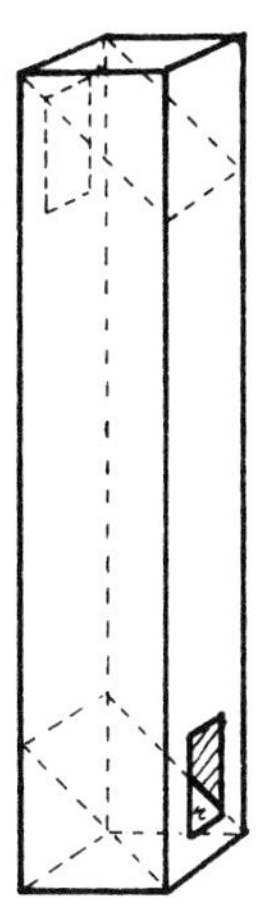

7 Fold the box back together to form the square tube shape and sellotape the edge to keep it together.

8 Now slide your mirrors through the slits you have made.
You can always sellotape these mirrors into position to prevent them from falling out.

What use could this be to you? What use could it be to small people or people in wheelchairs?

Extension
Now make a periscope that will allow you to look behind yourself. What do you notice about what you see?
The holes and slits are not in the same positions as in the present model, so think, design and measure, before trying to make it.

You might just hold up two mirrors to get the idea of angles, etc., before actually starting with your construction.

If you make a mistake the first time don't be put off; scientists and technologists often do this and we all learn from our mistakes.

What can be done to make what you see behind you be the right way up?

Here is a diagram of two sets of mirrors.

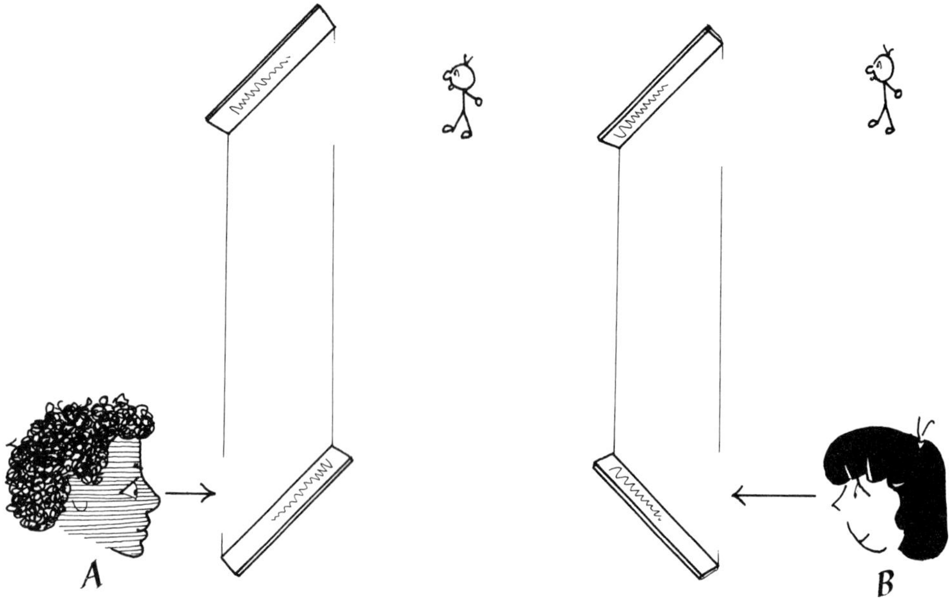

What would each of persons A and B see if they looked in the directions shown?

Which arrangement of mirrors would the submarine captain have?

SPIRALS

You will need
Strips of aluminium cooking foil
Length of cotton
Sellotape

What do do

Cut a length of aluminium cooking foil about 20 cms long and gently hold both ends and twist a few times.

Sellotape the 50 cms of cotton to one end and then sellotape the other end of the cotton to the top of an open door frame (or window frame, above a heated radiator).

What do you notice about the spiral?
What happens to the cotton?
What causes the spiral to twist?

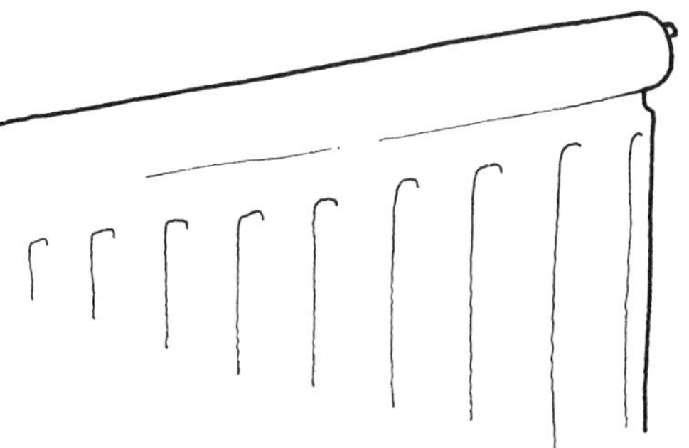

To look for

How many things have a twist in them? Here are a few, you look for more . . .
ropes . . . bean plants when growing up a stick, screws, drills, corkscrews, flumes
in the swimming baths, spiral staircase (does it matter which way the spiral staircase twists in an ancient castle?), springs.

Extension

PROPELLERS AND TWISTS

Take two pieces of stiff wire and twist them together to form a twisted rod of
wire.

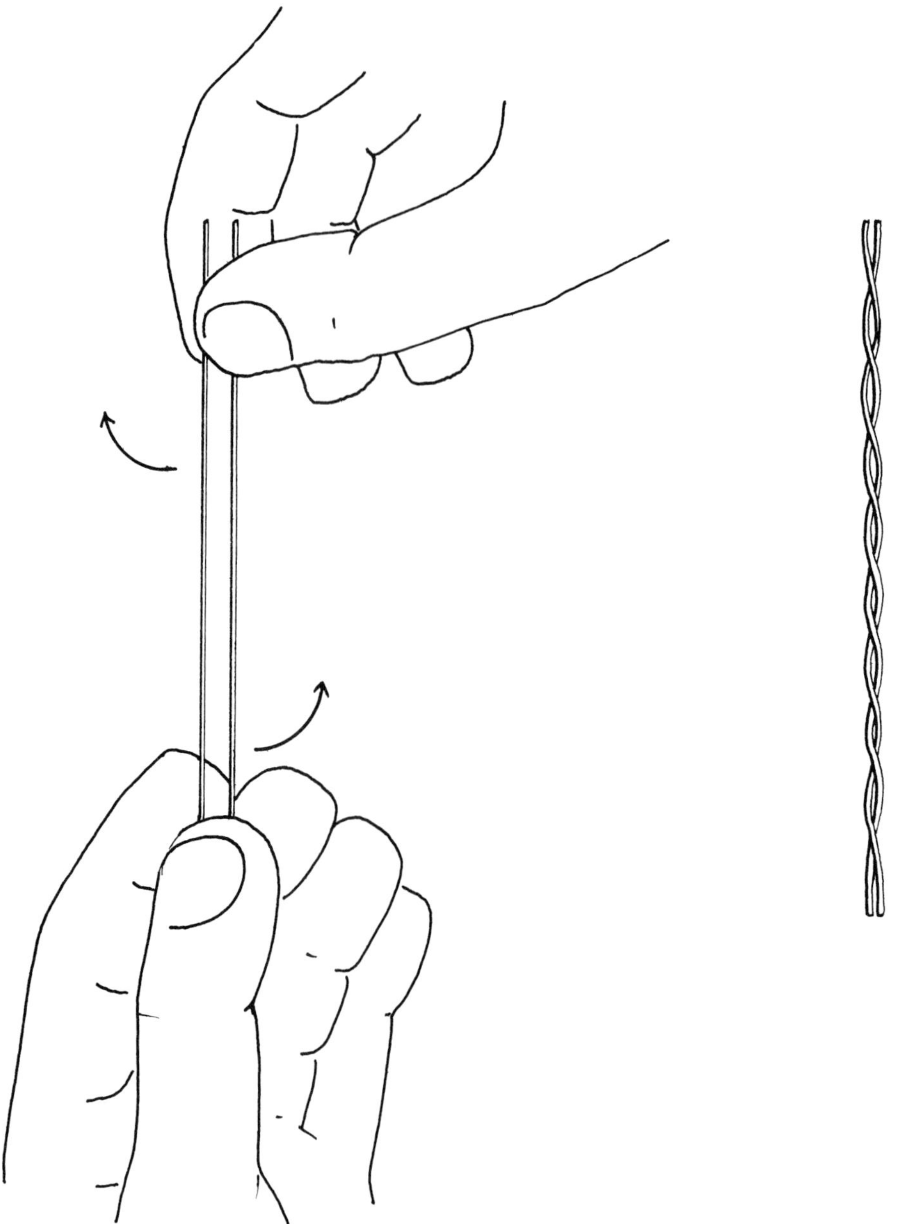

Make a propellor from plastic cut from the side of a strong icecream container.

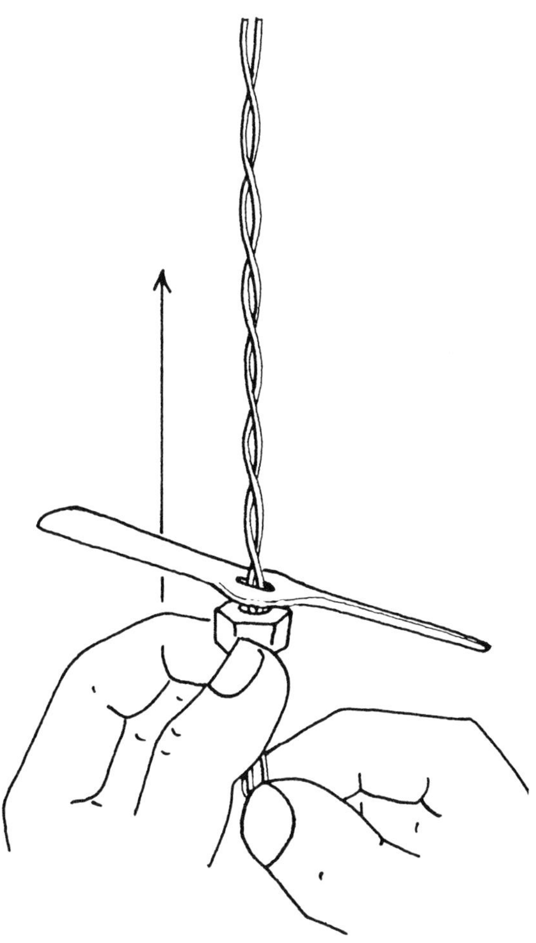

Put a slit in the centre just big enough to go smoothly over the twisted wire. Place a nut big enough to pass easily over the spiral without twisting under the propeller, and gently push the nut upwards on the twisted wire. The propeller should turn as it moves.

If the nut is pushed up quickly then the propeller will fly off the top and into the air.

PART 3

ARTICLES, REVIEWS AND CASE STUDIES

CONTENTS, PART 3

About Part 3

This part contains a few articles by people who have either expressed their own ideas or have put together a collection of ideas expressed in various articles.

They are designed for teachers and parents interested in knowing about some areas of teaching science and technology to a wide range of pupils with special educational needs.

Science and Technology in Special Schools
Paddy Rayner

Paddy Rayner

THE PROBLEMS

Coping with a wide range of special needs, Special Schools should be 'experts' in this field. In most educational areas they undoubtedly are, but when it comes to science and technology the problems are legion, particularly for Special Schools with secondary aged pupils.

Like most primary schools, the Specials have few qualified science teachers, no technicians, limited budgets and virtually no laboratory or workshop facilities. Curriculum planning is hampered by the lack of expertise of the staff, an irregular intake (particularly of secondary age range), lack of published materials specific to their needs, and internal demands on the timetable arising from the variety and degree of the pupils' handicaps.

Another crippling problem has been the isolation of Special Schools. Cut off from the mainstream flow of science/technology-based information and contacts, they have been unaware, until a few years ago, of what is available to them, even after attending some in-service training. John Taylor began turning the tide when working for the 'Insight into Science' team, and the North and South London Science Centres, the JISTT Unit and the Inspectorate all committed themselves to bringing special schools 'into the fold'.

PROBLEMS AND SOLVING THEM?

A total lack of science or technology qualifications in many schools was compensated by an interest in the natural and physical world and an awareness of children's delight in practical activities; these helped to bring about a change in the teaching styles. A move from 'book' and 'blackboard' science to practical work was helped by Gron Edwards from the Primary Science Centre. Armed with concrete science experiments and confronted by children with learning problems and/or unsettled behaviour, both staff and pupils gradually gained in confidence using the science activities. The pupils achieved small successes and became more motivated.

A curriculum structure was needed and 'Insight into Science' seemed to be the most useful. Two units were bought and piloted for a term, then expanded to all secondary groups and more units purchased. These were adapted where necessary, charting our own course through the units, and were taught initially as whole class practical sessions; but as pupil confidence increased, they began to use the cards in groups and work at their own pace. Much language and literacy work arose from these practical sessions, and recording work, both personal and display, improved remarkably as pupils began to take a pride in their work. All the secondary class teachers operated the scheme, after an initial piloting session of their own, working in tandem for half a term at a time.

There were, and there always will be, problems to overcome: pilfering, slap-dash approaches, reading comprehension and writing difficulties, safety awareness, pupils organising their own equipment, transporting resources from room to room. I could go on, but most of the needs of each child in the class were satisfied. A firm but encouraging approach was adopted by the teachers: successes were emphasised and praised and failures were played down.

A fairly large financial commitment over two years was made by any participating school to provide minimal resources for the scheme initially, but gaps necessitated much begging and borrowing, mainly from our very helpful local comprehensive school. Contacts often led to a formal link with CSE mode 3 general science work, based on modules, for pupils who had enjoyed their 'Insight' sessions and opted to take the course. This will continue under the GCSE arrangements.

Pupils attended the comprehensive school for practical sessions, follow-up lessons and homework being given at their home school. They were always accompanied by a teacher from the Special School, who joined in the teaching whenever possible. The link was never meant to be an exercise in total pupil integration, as the special pupils are taught in their own groups. All staff involved, from both schools, have expressed satisfaction with the scheme, and by their participation have had the opportunity to extend their knowledge of pupils with special needs. Both the pupils and staff have benefited enormously from this link and have increased in confidence, autonomy, knowledge and skills.

There are problems with such a link, the main one being the emotional demands on pupils having to return to the sort of environment in which most of them have met failure and unhappiness. Intellectual demands have proved daunting in some cases, but very slow working pupils have managed to speed up their rate of work with practice. A great many children in Special Schools suffer from high anxiety levels and, in spite of copious practice, often under-achieve in test situations, which adversely affects their examination results. However, the scheme has been able to provide pupils with a wide range of experiences, more than they could ever have received at their Special School, and has proved its worth time and time again.

The Education of Physically Handicapped Pupils of Secondary Age with Particular Reference to a Science and Technology Curriculum
David A. Coleman

In common with their colleagues in mainstream educational establishments, Special Schools have recently begun critically to examine and evaluate the curriculum they teach. They are now seeking answers to previously unasked questions concerning the relevance, in a changing world, of their traditional teaching styles, attitudes to and philosophy of, education.

This is particularly true for teachers working in schools for the physically handicapped, for here we have a population of school children who are denied the advantages of mainstream education not for reasons of impoverished or slow intellect, personality difficulties or for behavioural disorders, but because they possess, and are challenged by, a physical disorder which may require occasional medical attention.

It has been my observation that physically handicapped pupils consistently fail to do as well as their peers in comprehensive schools, even though, as previously stated, they are not a population selected by intellectual criteria. Statistically there should be a far greater number of congenitally handicapped students in our colleges, polytechnics and universities. Where are the disabled intellectuals with a fistful of 'O'-levels, CSEs and GCSEs? The answer is that they are few in number—so few, in fact, that their attainments guarantee local, if not national, news coverage. "BOY GAINS 9 'O'-LEVELS" is not a headline to sell newspapers, but, perversely, "COURAGEOUS WHEELCHAIR GIRL PASSES 'A'-LEVELS" is not only considered newsworthy but elicits positive public response (and, by implication, surprise).

Teachers of the physically handicapped are only too aware of their self-fulfilling prophecy inflicted upon their pupils. If the basic premise is that a physical disorder predisposes a child to require a specialised school environment, and if that environment, by its nature, is impoverished of those life experiences, essential resources and subject-teaching expertise, taken for granted by the able-bodied student, then the handicapped child will have the parameters of his or her development artificially restricted and thus appear to justify, by reason of academic failings and social immaturity, the original placement in an environment with a lowered expectation of performance.

As recently as the 1970s and early 1980s the typical curriculum in a school for the physically handicapped was hardly inspiring. Many members of staff would have been trained at a time when the emotional well-being of the child was pursued at the expense of striving for the realisation of academic potential. Keeping the children happy was considered a laudable goal, along with encouraging socially acceptable behaviour and teaching a stoical acceptance of a largely uncaring society. The timetable was particularly concerned with little more than basic skills supplemented with RE, music and, above all, craft.

Changes have fortunately been made now that the implications of such under-education of handicapped pupils has been realised. Many schools for the physically handicapped now have established GCSE courses, while others are vigorously pursuing programmes of integration with the local comprehensive school.

At this point I should like to nail my colours to the mast and state that it is my opinion that very many of the pupils at present being educated in schools for the physically handicapped ought not to be there, but should be integrated into mainstream schools. No Special School can possibly have the resources, the subject-teaching expertise or the normal social environment of a typical comprehensive school. When, at the age of 16, pupils from both schools leave to pursue their adult lives, it is obvious which one is better able to cope in society; I regret to say that it is invariably not the pupil from the Special School.

Two solutions exist which may be an answer to this problem. The Special School could try to become more like a typical secondary school in terms of physical, social, emotional and intellectual development of its pupils, or comprehensive schools should become truly comprehensive and open their doors to many more pupils at present in 'special' education.

The first solution would require a massive influx of cash, teacher re-training, curriculum development programmes and a complete change of teaching philosophy. The second solution would require much the same, plus a major rebuilding programme to correct problems of access in schools designed for use by the able-bodied. I ran a full integration programme at my present school which had 17 physically handicapped pupils fully involved and time-tabled in a normal secondary school. Of course there were occasional difficulties, but nothing which proved insurmountable, and after three years the differences in intellectual skill, social confidence, maturity and self-image between the integrated and non-integrated pupils were so great as to be a significant justification of the programme.

It would be naïve to suggest that *all* handicapped pupils should be integrated into ordinary secondary schools. Certain pupils have such severe physical and/or intellectual difficulties that involvement in mainstream education would be to the detriment of the specialised programmes they require. These pupils cannot successfully follow examination based courses or benefit totally from the academic or social possibilities of integration; yet, nevertheless, they should not be denied access to a varied and interesting curriculum which will motivate and stimulate as well as being of some practical use in helping them to cope with the life of enforced leisure which they will inevitably experience as handicapped adults.

In such a curriculum it is my opinion that it is essential to include Science and Technology.

What Technology and Science Can be Done?
David A. Coleman

Many physically handicapped pupils, because of their inability fully to explore their environment, and because of specific learning dysfunctions, have a severely restricted understanding of the properties of materials. Their spatial concepts are often distorted: in particular, pupils with spina bifida frequently can neither draw, nor interpret, three-dimensional diagrams; some cannot sort objects by selected criteria, nor do they fully understand such concepts as time, weight, volume or size. Many disabled children also have incomplete or immature language skills, both receptive and expressive. Such a pattern of learning problems will be recognised by the remedial teacher in normal school of those working with ESN (M) pupils. However, the implementation of remedial techniques meets with little speedy success in the school for the physically handicapped, since the learning problems are often compounded by severe physical limitations, including occasional problems of hearing loss, epilepsy, visual impairment and regular stays in hospital for corrective operations.

It is for such pupils that I have developed a science and technology programme. A curriculum developed pragmatically to meet the needs of each child becomes not only a remedial tool but a source of pleasure for the pupil and teacher alike. In my case, as a teacher trained in geography, the five years I have spent teaching science to physically handicapped pupils has been the most enjoyable and rewarding experience of my career.

Of course, for any activities to be undertaken, the difficulties imposed by physical restrictions have to be overcome. Many pupils will be in wheelchairs, others in calipers or standing frames, and yet others on prone trolleys or having one or more limbs in plaster casts. The child suffering from cerebral palsy will spend much of his time in spasm or, at least, have great difficulty in moving limbs to where he wants them to go. For all pupils to contribute equally in each lesson, much of the practical work is completed by other pupils and 'Sir' sitting or lying on the floor.

As stated earlier, it would be impossible for me to compete with the usual science/technology curriculum of a comprehensive school: I have neither the time, space, resources or expertise. Also, my science curriculum has to fulfil very different needs: it is not examination based, but must provide a practical knowledge of our world and present the child with experiences which will be of value in adult life.

My lessons, therefore, have to be a practical course, using readily available materials, which encourages the children to follow the usual scientific processes of questioning, testing, observing, recording, and drawing conclusions. To do so they must develop the social and linguistic skills which are taken for granted in able-bodied children, but so obviously lacking when our pupils first arrive in the secondary department of the school.

The equipment I use could easily be dismissed as junk. I have a policy of only using materials available in the pupils' homes or cheaply bought in local shops. Yoghurt cartons, string, garden canes, rubber bands, cornflakes boxes and cotton reels, are the backbone of my course. A whole range of such ephemera is stored (rather untidily) in my science area and is used by the children to construct various models and machines.

I begin each term with a problem and split up the pupils into 'design teams' who compete for a solution. The problem is usually (and deliberately) rather inane or humorously expressed and of apparently superficial importance, but uses processes and principles which, I hope, they will see to be of importance and relevance at the end of the activity. Examples used in the past are 'How can I lift a hot cup of coffee off my desk without using any part of my body and without the coffee spilling?' 'Design a machine which will carry a marble balanced on a cotton reel from one table to another one metre away'. 'Build a car which will climb up a 25 degree incline carrying a glass of water' (which must not spill). Such problems usually prompt great discussion, and before we divide into our working groups many ideas and possible difficulties have been thought of, giving much for the pupils to test and make decisions about.

Each term we keep a large wall book—called our 'Eureka File'—in which any ingenious discoveries are recorded (this saves time in later terms by avoiding the problem, say, of 're-inventing the wheel'). Each team produces a working diagram which is mounted on a display; a diagram of the finished model is later produced so that the two may be compared and any necessary modifications discussed by the whole group. The last two sessions each term are reserved for the big competitions where the models are put through their paces.

Points are awarded for success, adherence to the rules of the problem, design and innovation, and originality. Models are then put on display for a short while before being tested to destruction or broken up for their parts.

The value of such an approach is obvious to me as I watch the growing excitement, the motivation and improved linguistic, social and manipulative skills of the pupils, but the real test is whether there is a carry-over of method and skill into areas of science. The topic by this time can be seen to have relevance and many new skills have been developed.

The final term each year has a 'taught' science course which is related to the practical work of the previous two terms. Following the vehicle building terms, for instance, I teach the pupils about pulleys, levers, gears and structures, and indeed they are all better able to handle the language and understand the concepts than would have been possible without the earlier practical work and problem solving activities.

Year two is concerned with Living and Growing (plant life) and with Human Biology and Health Education.

In year three we explore the world of chemicals—in our food, chromatography, fuels, etc., and finally in years four and five I repeat a Technology course and the Human Biology course but now at a much higher level than in earlier years.

Many areas require much improvement, but the value of a course based on pupil need, rather than examination syllabus, is that such change is possible. I can also evaluate the pupils' progress and so help them *reach their highest potential*—a fact which is the highest single factor in developing their positive feelings of self worth. None of my pupils could pass high grades of GCSE, but this does not prevent them thinking and producing novel solutions to problems. With the knowledge gained in the science and technology classes, all are much better equipped to handle and understand the properties of everyday materials. A complex world is less confusing for them and, above all, they have such an improved self image they would not readily avoid the necessities of daily problem-solving which form part of a typical adult's life.

Science and Technology for Pupils with Limited Learning Abilities
Alan V. Jones

There are a number of teachers, already teaching science to the slow learners in comprehensive schools, who would not have to change their style or material appreciably to teach the mentally handicapped child, should such pupils be integrated into their school.

This title 'mentally handicapped' is not one that is now widely used, but the term 'very slow learner', or 'child with limited learning ability', is generally used and such pupils are to be found either in special education or integrated into the comprehensive school and often located in slow learning groups (bands, streams or strands), sometimes called the retarded or restricted learner group.

Any teacher must know the individual pupils' characteristics in any class he or she teaches, to ensure effective learning for all the pupils. The variety of characteristics within the slow learning group must also be accompanied by a variety of suitable teaching methods. Mentally retarded children, from the many causes, have some general characteristics for which similar teaching methods can be selected. For example, many children who are slow learners are often poor readers, with limited language skills, weak in mathematical skills and not able to understand abstract concepts. Their attention span in class can be quite short and sometimes they are attention seeking in one way or another. Often they have limited manual dexterity. From these facts it can be seen that a written work sheet with some numerical bias would not be the most suitable way of teaching science to these pupils. Neither could any system which requires detailed reporting or examinations be matched to their abilities and needs, but it has often amazed visitors to see how the children can be taught to undertake quite detailed science and technology experimentation by verbal explanations and using suitable methods managed by a skilled teacher (sometimes with classroom helpers). Although the pupils have limited skills, they must be taught to develop these skills to a maximum level and this means not neglecting the use of the written word and expecting them to write *something* down (if at all possible).

Sometimes the verbal reporting back of the findings of an activity is sufficient, or the making of a piece of apparatus is an alternative means of communicating the results. The grading of such a piece of work must reflect the personal effort of the individual and not be scored F on the class norms (if integrated into a normal class). Such pupils usually have a low self image and so encouragement is needed whenever possible: they should not be given deflating F grades, but encouraging comments to help them strive for higher things. Personal profiling work and criteria referencing can effecively show the development of such pupils.

When giving instruction—verbally, in writing, or on a tape recorder—it is often best to keep things simple and in short, manageable doses. Let one task be completed before the next is given.

SCIENCE AND TECHNOLOGY FOR ALL

When science/technology is said to be for all pupils, it does not mean the *same* activities for all pupils but *some* science/technology for *all* pupils. The tasks and materials selected as the themes for the whole class must be individually selected for the retarded pupil so that the material matches his or her abilities and skills. On some occasions the pupils become fascinated by a particular experiment and want to repeat it over and over again. This process may amaze the teacher who often finds the attention span short. If this situation arises, introduce slight variations to the experiment to keep up the interest level and extend the content being taught. The retarded child is often able to deal effectively with concrete material, experimentation and direct observations rather than with mentally manipulating ideas. It is not necessary for retarded children to gain a series of solid, academically logical scientific foundations, as they often cannot remember what they did the day or week before, but it is important that each activity develops some suitable skill (manipulation, construction, dexterity, writing, observing, reporting, measuring, accuracy, mathematics skills, etc.).

It is highly probable that the retarded child will not mentally progress above a mental age of his/her early teens (if that), and so, whatever topic is being covered, suitable 'concrete' activities must be sought.

Science and Technology for the Auditory Handicapped Child
Alan V. Jones

A child suffering from partial or total deafness has to develop an effective communication system with a hearing person and meaningful speech is often only acquired after long and patient practice and schooling. A hearing person does not appreciate how much information is acquired unconsciously through hearing and listening to others. Thought and reasoning, too, are often carried out in the mind in the form of word patterns and sounds, but a deaf child is deprived of word sounds and formation and so cognitive development is often hindered. To succeed in a hearing world the deaf child has to develop patterns of reasoning and a communication system that is acceptable in the normal world. Piaget showed that broad concept development often depends upon *all* the sensory experiments working in total harmony and it has been shown that hearing impaired pupils help to compensate for this by doing a lot of active experimenting and other interactive situations. Research in the USA, based on a series of investigations with a large number of students, showed that the abstract mental abilities of the deaf child were significantly inferior to their hearing peers, and that the 'conceptual categorisation skills of deaf children can be substantially increased through the use of specific experiences in clarification and physical manipulation of objects'. It has been shown that practical science can help the formulation of abstract concepts and indirectly aid communication skills. Harry G. Lang, himself deaf, commented that 'a review of the literature reveals no major theories of instruction for deaf children', and that in the area of science education the 'various strategies and paedalogical techniques that have been reported upon in the past' are no more than 'individual stabs in the dark'. The lag in language development presents three common problem areas in the physics classroom—syntactic, lexicon and experimental. Syntax is related to the ability of hearing impaired students to understand what is being expressed in the sentences, as deaf children often confuse the subject and object in a sentence. Lexicon refers to vocabulary and the often hidden subtleties of the English language. The language lag does not necessarily mean that the child will never be able to make up the deficiency or to develop cognitive abilities.

Lang continues, 'Deafness as a loss of hearing is a disability with secondary effects on language development, socialisation and cognitive growth and 'handicapping' conditions which result from society's treatment of the deaf child. The extent to which these conditions become handicaps depends on personal and societal expectations.'

It was found that the deaf child integrated into a normal school is often reluctant to admit that she or he does not understand and consequently heaps up further problems; an individualised learning sequence may help the child overcome these problems. Many of the curriculum projects developed in the USA have given hints to teachers to help them cope with various groups of handicapped pupils and also some teachers have published individualised learning schemes based on these projects at the primary and secondary school level. This is true of the hearing impaired as well as other disabilities. It should be realised that science is not a totally fixed set of subjects, topic or rigid experiments but is a source of knowledge which can be approached in as flexible a manner as possible in order to convey the concepts effectively. This can mean the adaptation of suitable experiments using a 'sellotape and elastic band' approach, or perhaps contacting another school where there are other handicapped pupils (whether special or ordinary school). Many handicapped pupils, sometimes with time on their hands at home, like to repeat interesting work done at school; this could apply to simple science, technology and nature study, and to back-up reading.

Teaching Science and Technology to Deaf Children
Richard Wheat

INTRODUCTION

Deafness can occur in many forms and in varying degrees. Children found in special schools for the deaf may be either profoundly deaf, with a hearing loss in excess of 90dB, or partially hearing, with a hearing loss often in the 60–90dB range.

The main criteria for giving a child a place in a special school for the deaf are based on the educational needs for each individual child. Children born profoundly deaf have not had the opportunity to acquire language or speech naturally, and therefore the child usually lacks adequate receptive and expressive language. Congenital or early deafness creates a chain reaction in the learning process which blocks the normal development of language, speech, reading, abstract thought and socialisation.

The spoken language is one of our most efficient means of communication; language is also the main and most efficient 'tool' for the brain to think with. Without language we cannot communicate efficiently with others or think and talk to ourselves in a normal manner.

In a baby, it is thought the brain has a special facility for learning language; by the age of three or four this facility has declined. Too often cases of deafness are not diagnosed early enough and remedial action not begun until the age of three or even later. Early diagnosis is vital to enable the child to take advantage of the special facility for learning language.

The degree of success a deaf person has in acquiring language depends on the severity of the hearing loss, the person's intelligence, the age of diagnosis, plus the factors which affect the development of all children—supportive home background, etc.

Pre-lingually deaf children find difficulty in understanding spoken language even when using a high powered hearing aid system and the written word. In addition, written and spoken versions of English may not be identical. A deaf child may well understand nouns, active verbs and some adverbs, but conjunctions, prepositions, pronouns and tenses often baffle or seem superfluous to the deaf child.

Often a school's teaching philosophy is based on the oral method, teaching speech, and speech reading, together with sign-supported English when considered relevant and necessary.

THE TEACHING OF SCIENCE AND TECHNOLOGY AND ITS DEPENDENCE ON
LANGUAGE DEVELOPMENT

When teaching science and technology to deaf pupils five main factors have to be kept in mind:

1   There is often the usual wide range of intelligence within a group of deaf pupils.
2   There is a wide range of mental, manipulative and practical abilities in any

so-called homogeneous group of pupils. The frequency of deafness in the general population is about one in a thousand.

3   It is necessary to understand the concept level of the pupils and to start from concrete materials and then move slowly towards abstract concepts.
4   One of the main (or indirect) aims of all lessons is the development of the child's language.
5   The child's environment often lacks the stimulation provided by the sound of voices. For the deaf child information arrives in mainly visual form, with incomplete auditory patterns.

Science can be a useful stimulation to encourage language development. During a typical lesson, written and verbal instructions concerning the experiment or investigation are usually given. These instructions must be understood if the child is to act upon them and the experiment successfully carried out. When giving instructions, sentences must be kept as simple as possible, preferably with one verb and few prepositions or conjunctions. Even a simple noun or verb may have to be explained, as will the meaning of whole phrases, e.g. 'Put one crystal in a beaker full of water'.

In the above sentence, depending on the level of language attainment, the word 'put' might have to be explained, along with the word 'in' as opposed to 'on', or 'under', or 'behind'. Assuming the pupils know the word 'beaker' from previous science lessons, the phrase 'beaker of water' or 'beaker full of water' may not be understood.

By the time you have explained all the words in a passage, either the lesson is almost over or the main point of it has been forgotten. A happy balance must be achieved between language teaching, science teaching, practical work and maintaining interest.

As much language as possible should come from the child, as a result of various activities and the desire to do more things, together with asking questions and advice; the teacher should build upon and extend this contact and interest.

With much basic language to be taught, the teacher must always ask himself, is the science material relevant to life? Is the topic really what the pupils want or need to know? How can the lessons be made more relevant to everyday situations?

'The problem' of the science teacher of the deaf is not one of being heard, but of being understood (not an uncommon problem for any teacher, you may say).

GIVING INSTRUCTIONS

Before practical work can commence it is important that the pupils understand what they have to do; it is difficult to stop them once they have moved to all four corners of the laboratory, although their attention can be gained visually.

When explaining the work the teacher has to ask many questions, and the same question must be repeated many times to different pupils, because for one pupil to understand does not mean they have all 'heard' and understood. Simple questions, such as, 'Where is the thermometer?' and 'Show me the beaker', are often asked in an attempt to encourage speech and to ensure the pupils are following what is being said.

Once a pupil has completed an experiment or investigation there is a good opportunity for the communication of results to come from the child either in the form of orally answered questions or of written work. Some questions, although concerning the science work, need not be aimed totally at improving scientific knowledge, but should also be designed to improve the language and communication skills.

WHAT TYPE OF SCIENCE AND TECHNOLOGY?

The work must be at what Piaget called the concrete level, ie. there must be something physically in front of the pupils for them to relate to easily and directly. The work tasks must be of short duration to allow for short concentration spans, as the child may be fatigued by his or her lack of sufficient language with which the brain can operate efficiently. Revision and reinforcement of language are some of the main concerns from lesson to lesson: science activities can introduce new language and, more importantly, give the children opportunity to use it meaningfully.

Science lessons for the deaf should be built around common materials that are readily available and can be easily handled. Pictures can be helpful but are not as good as actually doing something themselves; the more tangible an item is, the more impact the language has in the lesson.

Science activities for deaf pupils might include some common experiences that hearing pupils take for granted. A hearing child, having the benefit of hearing speech, constantly learns things unconsciously. Some very elementary biology lessons come into this category and are almost common sense to a normal child, but even the major parts of the body, their use and function, are often not easily understood by the deaf child.

Biological Activities for Visually Impaired Pupils
Ron Hinton

The less-severe types of visual impairment, where print is still readable and many observations can be made in the usual way, need not concern us here except to point out that safety factors are important in fieldwork and where bunsen burners and other potentially dangerous items are in use. This is not to suggest that these and other scientific and technological activities should be avoided, but that procedures should be adopted for their safe use.

It must also be emphasised that the illumination of books and work surfaces needs to be adequate. Work by Dean Southall of Loughborough University showed that, although with the fully-sighted individual poor lighting does not cause much reduction in efficiency until it is really bad, a slight reduction in the level of illumination for a student with a visual impairment can cause a dramatic reduction in performance.

It is important to provide practical activities for the blind student, of sufficient quality to enable him to establish scientific concepts on a sound basis of experience. Many of the pitfalls of misunderstanding which beset ordinary students, such as those of size relationships and scale, are all too easy to fall into if you cannot see. To teach a biology/science/technology syllabus in a second-hand way, through rote-learning, is restricting to any student, but to the blind student it can be particularly damaging. A little extra trouble taken to provide practical experience is worthwhile. Where live organisms are not accessible, museum specimens, models and tactile diagrams can be employed to fill in the gaps, as long as these are carefully related to the living organism.

Bear in mind also that the visually impaired student is unable to make the sweeping overview which, for the fully-sighted observer, gives cohesion to his observations in any larger study area. The visually impaired student will make observations, by touch or by vision at close quarters, which may be disjointed by gulfs of space and time from his other observations in the same area. Tactile diagrams or other graphic displays provided by the teacher, or built up by the student from his own field data, may help to show the relationships of the individual observations to each other. The student can make observations, refer these to a plan or display, and then use this to co-ordinate further observations. Methods of producing such quick tactile displays include the Sewell drawing board and biro, the Braille apparatus, and the spur wheel. All of these are available from the RNIB.

Systematic fieldwork is perfectly feasible for blind students provided that the study area has been reconnoitred by the teacher beforehand; possible major hazards of terrain should be avoided or the students provided with protection. Areas should of course be appropriate to the age and experience of the students. Blind students' mobility training will enable them to cope with minor hazards such as overhanging branches, but forewarning of these will naturally be part of the briefing before going into the field.

The teaming-up of visually-handicapped students with fully-sighted colleagues in integrated classes, of course, solves many of the problems of fieldwork, provided that the visually-handicapped student remains a genuinely active and exploring member of the team, and does not become a passive recorder of the observations of his sighted peers. The assistance of a sighted observer in the shape of a teacher or classmate is unavoidable in some aspects of the teaching of sciences to visually impaired students, but where a viable alternative exists or can be devised, then it should be adopted.

When the visually impaired student needs to make quantitative observations in the field, much of the standard apparatus is manageable. For example, quadrat frames for sampling vegetation are successfully used, either laid systematically across a study area or thrown by the usual random sampling technique. In the latter case a light anchorage string to enable the student to locate and retrieve the frame has been found helpful. Where the larger quadrats are used, additional subdivisions by elastic bands or tapes will give the student extra reference points for accurate estimation. In fact, in any estimating task the blind student should always be provided with enough points-of-reference or base measurements to support a valid estimate. The commercial point-sampler quadrats are particularly easy for a blind student to use because the vegetation to be examined is precisely located by the sampling points in normal use.

There are a number of tape measures and rulers with tactile markings available from the RNIB, and also such items as protractors and the Braille-marked Silva navigating compass may be useful for surveying activities. However, it is easy to make a simple measuring rod with notched or studded markings to any appropriate scale, from wood, or from plastic or aluminium curtain rail, and we have found a knotted cord line cheap and easy to use (see below).

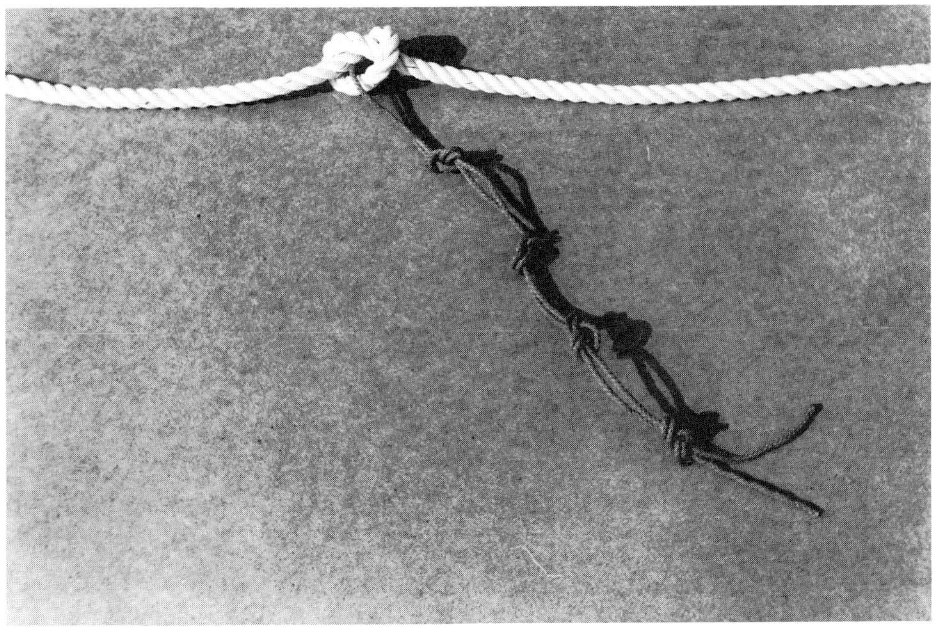

The one illustrated has added marking cords like the tags of a surveyor's chain and is convenient for traverse measurements along a string baseline, but a simple regular knot at, say, 5 cm or half-a-metre may be quite sufficient, and easier to use.

Most ecological projects involve identification of the organisms found at some stage. Although the transcription of standard identification keys into Braille or onto audio tape may not be sensible because the identification criteria are predominantly visual ones, there are many other criteria (normally 'redundant') which can be used instead, as long as the circumstances in which they are to be used are specified to avoid confusion. The present writer has had success with a key to British Hardwood Trees in Braille. This was specially devised for blind students and differentiates the species by means of leaf characters detectable by touch. The key takes the form of a pack of cards sorted by means of marginal punched holes in the manner of some library indices. It is intended for broad-leaved trees growing in more-or-less wild conditions anywhere in Britain. Similar keys have been produced for other well-defined morphological groups, or for specific habitats.

Just as with the fully-sighted student, any key of this type needs to be introduced with bench-top specimens first so that students become familiar with the manipulation of the cards, and with the necessary decisions which need to be made in examining specimens. The aim will be to become so familiar with the specimens after further work in the field that the key can be largely laid aside as the student works from memory, preferably incorporating characters of his own devising to make identifications more secure, as he becomes more familiar with the plants.

Handicapped Students in Integrated Situations
Alan V. Jones

SOME BASIC HINTS

The following points may be helpful if you have handicapped pupils in your class or about to arrive in your school.

1   The fact that he or she has approached the school means that a lot of groundwork has already been done, ie. consultation with parents, previous teachers and inspectors and the pupil must have been recommended as suitable for integration.

2   Before the beginning of the first term the child should visit the school and check all the facilities. This is particularly important for the partially sighted and wheelchair-bound pupils.

3   Check if special apparatus, equipment or medical and toilet facilities are needed, and also what to do in case of an accident or a breakdown in the child's daily health. Have the phone numbers readily available for people to contact. Accessibility to toilets and dining-rooms is as important as the accessibility to classrooms and laboratories. Children who are incontinent might need a small changing room close to the toilet area; some schools have converted one toilet cubicle for this purpose.

4   Run over with the pupils the procedures to be followed in case of fire or fire practices. Ensure that a place near the classroom door is set for the wheelchair bound child and in case of fire ensure that a child or teacher is responsible for the quick exits.

5   If possible, when writing timetables and booking rooms, check that these are as near the exits (or entrances) as possible.

6   Invite a few of the other classmates to this preliminary meeting if appropriate, and also the ancillary staff assigned to the school to look after the handicapped pupils.

7   Ensure that academic staff, cafeteria staff, librarians and caretaker know of the presence of the child.

8   All teachers in contact with the child should know the extent of the handicap and know what to do should immediate medical care be needed.

9   For laboratory work check height of tables, benches, accessibility of water and gas taps, and see if the child is able to work alone or as a member of a group. (Nonslip plastic sheets for the bench are useful, and for some pupils enlarged handgrips are beneficial—contact Nottingham Educational Supplies, Melton Road, Nottingham.)

10   If the child arrives by special car or taxi, ensure it arrives at the most appropriate doorway for easy entry.

11   If visits or outings are made, it is important to make the effort to take the handicapped child also and explain this to the contact at the place to be visited. Ensure that the 'taxi' service which buses the pupils to school knows about the outing so that any delay in the return is known.

12   If the child has muscular or neuromuscular complaints and has speech or movement defects, prepare the classmates initially so that they can be as supportive as possible. Encourage pupils with speech defects to make verbal contributions as practice can improve confidence and communication.

13   Some pupils might only be able to communicate or write with typewriters, microwriters or other aids, and although these do make some noise, they are not usually disruptive. Some pupils might ask to use tape recorders or have small amplifiers to help their hearing, or may need the teachers to wear a small lapel microphone.

14   When talking, face the class (this is particularly important for the children who need to lipread or for partially sighted pupils).

15   If new words or technical words are used, give these prominence during the lesson or give a handout. (In the USA sign language interpreters accompany any deaf child integrated into the normal classroom.)

16   When talking to deaf or partially sighted children, stand still and try not to stand in front of a bright light or window. Try not to turn your back on the class when important points are made.

17   If a film or video tape is used, give a summary of the sound for the deaf child.

18   If you change rooms for a lesson, etc., leave a Braille message or message imprinted on a soft surface (a piece of ceiling tile or polystyrene) in a prearranged place for the blind pupils. For a blind child tactile diagrams which can be felt with the fingers might be needed. These must be two or three times the size of normal visual diagrams as the fingers are not as sensitive as the eyes. If diagrams are used with a class containing a blind child, describe them fully to allow the child to get a mental picture.

19   Instructions for experiments for the blind child, or for any children with learning difficulties, can be read on to a sound tape; alternatively a synchrofax machine can be used.

20   Specific advice on the education of any particular child can be obtained from the special education adviser or from the associations related to the particular handicap, eg. Spastics Society, Muscular Dystrophy Society. Usually the adviser, child or parent knows the relevant addresses. Ask for advice from the local Special School.

21   Alternatives to written work can be typed messages, microwriter printouts or tape recordings. In all circumstances expect the child to be as normal as possible and to do the tasks assigned to the whole class. Sometimes well-meaning, over-protective parents, friends, teachers and ancillary helpers overcompensate for the child's disabilities. The child *must* be encouraged to cope in the real world. His abilities need to be emphasised, not his disabilities.

22   You might find that some children, particularly those paralysed from the chest down, have perceptual problems. They often find difficulty in perceiving diagrams in three dimensions and also confuse up/down, back/forward and other

comparative phrases. They use these words frequently in their conversations but have difficulty in knowing which is which. This might be due to their lack of 3D feelings (if they are not able to feel things below a certain height then it is difficult to feel what is down). The child's report from the previous school or educational testing service or psychologist should mention this.

BIBLIOGRAPHY AND USEFUL BOOKS

Science for Handicapped Children (1983), Alan V. Jones. Souvenir Press, London. ISBN 0 285 64969 8.

Science Education and the Physically Handicapped (1979), H. H. Hofman and K. S. Ricker (eds.). National Science Teachers' Association, Washington, USA.

Teaching Chemistry to Physically Handicapped Students (1981), American Chemical Society, 1155, 16th Street NW, Washington DC20036, USA.

Able Scientist—Disabled Persons, Careers in the Sciences (1984), S. Phyllis Stearner. Foundation for Science and the Handicapped Inc., USA.

Teaching Handicapped Students Science, Marshall E. Corrick (ed.). National Education Association, USA. ISBN 0 916655 00 8.

Mainstream Teaching of the Physically Handicapped in Science. A Resource Book (1985), E. C. Keller, T. K. Pauley, E. Starcher, M. Ellsworth, B. Proctor. Printech, 1125 University Avenue, Morgantown, WV 26505, USA.

Testing Physically Handicapped Students in Science, D. R. Brown, E. C. Keller, H. G. Lang, K. S. Ricker. Printech, USA.

Integration in Action, S. Hegarty, K. Pocklington, D. Lucas. NFER/Nelson, Windsor. ISBN 0 85633 238 0.

Schools, Pupils and Special Educational Needs, D. Galloway. Croom Helm, London. ISBN 0 7099 1175 0.

Teaching Science and Mathematics to the Blind. Royal National Institute for the Blind, 224 Great Portland Street, London W1.

The Hearing Impaired Child in the Ordinary School, A. Webster and J. Ellwood. Croom Helm, London. ISBN 0 7099 3630 3.

How to Reach the Hard to Teach, Paul Wildlake. Open University Press, Milton Keynes. ISBN 0 355 10194 1.

Yes They Can (1974), K. J. Weber. A practical guide for the adolescent slow learner. Open University Press, Milton Keynes. ISBN 0 355 00244 7.

The Teacher is the Key (1982), K. Weber. A practical guide for teaching adolescents with learning difficulties. Open University Press, Milton Keynes. ISBN 0 355 10047 3.

Educating Hearing Impaired Children (1984), Michael Reed. Open University Press, Milton Keynes. ISBN 0 355 10422 3.

Research and Development Concerning Integration of Handicapped Pupils into the Ordinary School System (1980). National Swedish Board of Education, Stockholm.

Directory of Non-medical Research relating to Handicapped People, Jim Sanhhu. Handicapped Persons Research Unit, Newcastle-upon-Tyne Polytechnic, 1 Coach Lane, Newcastle-upon-Tyne.

INDEX